101 Analogies

For use in Psychotherapy

DAVID SEATON, PhD

101 Analogies for use in Psychotherapy

Authors note: All references to clients in this book come from my clinical experience over the past 20 years of conducting therapy. Names, places, and other identifying details have been changed and altered to protect the privacy and anonymity of the composites of a number of clients who share similar issues and are equally protected with name and information changes to remain confidential. Any similarity between the names and stories of individuals described in this book and individuals known to readers is coincidental and not intentional.

First printing 2011

ISBN 978-0-578-04926-7

Table of Contents

Introduction

After successfully traversing the various trials and tribulations of graduate school, the neophyte clinician leaves with a vaunted and potent armory of theoretical knowledge just waiting to be efficiently disseminated to his or her willing clientele. Armed with this complex dialect steeped in hallowed tradition, the clinician begins to formulate elaborate and useful interpretations that will undoubtedly provide the classic "Ah Ha!" experience in their eagerly awaiting patients. After all, we have been speaking this language to impress our peers and family for the last two to four years and have done so quite nicely, thank you.

However, as most of us experience, a majority of our clients rarely fully grasp or appreciate these pearls of arcane wisdom. Sure, there is the rare patient that almost demands this type of interchange. Perhaps to justify the fee they are paying; or to continue in their intellectualized defense, moving them farther and farther away from deeper emotional solutions. Instead, these theoretically laden interpretations, while probably accurate, are

1

often met with feigned acceptance, probable indifference, or even hostile rejection. On the drive home, the ambivalent spouse, reluctantly dragged in to the first visit vowing never to return, may label this type of interaction as "high falutin' gobbledygook," or "psychobabble."

Perhaps nowhere is the discrepancy of understanding between therapist and patient better illustrated than in the hilarious interchange between characters portrayed by Billy Crystal and Robert De Niro in Universal Pictures' *Analyze This*. In this scene the hurried and overly pedantic psychiatrist (Crystal as psychiatrist Dr. Sobel.) attempts to analyze the origin of the increasingly perplexed and hostile mafia boss' (De Niro's Paul Vitti) panic attacks and erectile dysfunction by invoking that most hallowed of all psychological interpretations; the Oedipal Complex.

Dr. Sobel inquiring as to the relationship between Paul Vitti and his father: "So you weren't getting along then?"

Vitti: "I was in a kid gang, mostly hooligan shit."

Dr. Sobel: "He didn't like that? He didn't approve?"

Vitti: "No he didn't approve. He slapped me around a couple of times."

Sobel: "Then what?"

Vitti: "Then he died."

Sobel: "How did that make you feel?"

Vitti: (Sarcastically) "It felt great. It was a wonderful feeling. How'd that make you feel."

Sobel: "Think about it were you angry… were you afraid… sad?"

Vitti: "Maybe all of those things."

Sobel: "Any feelings of guilt?"

Vitti: "About what? I didn't kill him."

Sobel: "I know that but I'm just speculating Paul, that maybe in some way maybe you wanted him to die."

Vitti: "What?! Why would I want my father to die?"

Sobel: "You said that you were fighting, he slapped you around. You were rebelling against his authority. There may have been some unresolved Oedipal conflict."

Vitti: "English, English."

Sobel: "Oedipus was a Greek king who killed his father and married his mother."

Vitti: "F****ng Greeks."

Sobel: "It's an instinctual developmental drive. The young boy wants to replace his father so he can totally possess his mother."

Vitti: (incredulous) "Wait a second. You're saying that I wanted to f**k my mother?"

Sobel: "It's a primal fantasy."

Vitti: "Did you ever see my mother?"

Sobel: "Paul…"

Vitti: (Becoming enraged) "Are you out of your f#*#ing mind?"

Sobel: "It's Freud!"

Vitti: "That Freud was a sick f**k and you are too for bringing it up…. Blech!"

What is the therapist to do? Forget the rigorous training received during countless hours in myriad clinical settings? Forsake ones devotion to a long held belief in a theoretical orientation? The answer is a resounding no, but we might want to save a majority of these sophisticated interpretations for cocktail parties or study groups. Instead, we must search for a transitional language that allows the clinician to convey his deep understanding of human behavior to the client in an efficient and mutually understood manner. There is a need for a common language that seeks to join therapist and patient in a more meaningful therapeutic rapport.

Having run into this particular therapeutic quagmire, I notice analogies spontaneously presenting themselves that allow me to couch interpretive dialogue in a more easily understood format. When this occurs, there is often an immediate change in the client's perspective. They can now experience a momentary emotional distance from their problem. This new perspective allows them to view their situation in encapsulated form, having a clear beginning, middle, and end. These interchanges also allow the patient to feel that the clinician relates to their problem on a more human level. It engenders trust and deepens therapeutic rapport, allowing the client to subjectively sense the clinician's ability to quickly understand and analyze large chunks of their current situation. At the same time, it allays the client's fears of their psychotherapist becoming overwhelmed by the sometimes intense emotional content of the situation.

I am quite sure this is an ability I share with other clinicians. Still the power of this simple linguistic technique continues to surprise me in how it aids the therapeutic process. Invariably during termination sessions patients site various analogies as particularly useful in solving their presenting complaints. They report that this fresh perspective was often the turning point in their therapy, allowing them for the first time to form practical solutions. I must note that this was somewhat embarrassingly in contrast with what I thought was their significant therapeutic breakthrough (mine usually centered on some brilliantly formed historical interpretation concerning object relations theory). But alas, ours is a business that hinges on the "customer is always right" philosophy. The ever-encroaching role of managed care and its mandate to utilize solution-based approaches promotes the use of a simplified goal-oriented language. In this way, analogies have also provided a systematic way to structure goals in short-term psychotherapy.

I find that analogies are extremely helpful in my work with men, especially those that are brought in by others, and who view therapy as akin to having a limb removed sans anesthesia, or the equally painful "day at the mall". My theory is that these reluctant participants may feel threatened by the standard therapeutic language in which their wives and overly educated male therapist are much more adroit. Analogies serve to disarm these fears and surprisingly open them up to the use of more therapeutically laden dialogue deeper in the therapy process.

The purpose of this book is to offer the clinician a compilation of useful therapy-tested analogies accumulated over the last 20 years of the authors experience in private practice. I hypothesize that most of the analogies are of my own spontaneous creation. However, I have been blessed to work with a myriad of talented professors, supervisors, colleagues, and even clients, so I may have lifted an unconscious idea or two along the way.

The analogies are categorized by problem, with a brief description of their therapeutic utility. As a resource guide, I believe this book can benefit the beginning clinician looking to convey theoretically derived knowledge in a practical and commonly understood language across a wide array of therapeutic situations and clinical populations. For the more seasoned clinician some of the provided analogies are an addition to their already extensive repertoire of clinical interventions. As you examine these informative and entertaining analogies, some will intuitively make perfect sense and you will be itching to implement them in the therapeutic setting. Others will have you scratching your head regarding the usefulness of the metaphor. Therein lays the challenging quality of conducting any psychotherapeutic endeavor; that is individual differences. Where one finds a particular analogy extremely helpful, another client will find it useless (don't worry, your client's unproductive and awkward silence followed by the facial expression one gets when viewing some forms of abstract art will help you discern which is which).

For example, one analogy I often use in marital therapy compares the process of sustaining and maintaining a healthy marriage to that of receiving an original '63 Corvette on that same wonderful day. The analogy posits that while this car is not in pristine condition it is in good working order, contains all original parts, and has the potential to be restored to mint condition, thus becoming a sound investment worth much more than when initially received. Does one simply do the required maintenance that keeps it in its originally received condition? Does one store it in the backyard vowing to make the time to restore it to mint condition instead becoming distracted by other essential day-to-day concerns (career, kids, alcohol, golf) thus allowing grass to grow high around the ever rusting classic?

On the other hand, does the couple make the necessary commitment to continually restore the car to showroom condition? To me this is obviously a first-rate way to assess the current state of the marriage of a couple I might be seeing in psychotherapy. At the same time we can conceptualize the corresponding amount of work the couple will have to undertake to progress to the desired state of marital functioning. However after painstakingly relaying this analogy for editorial approval, my girlfriend innocently stated, "But I would want a new car".

In addition to the book's use as a resource guide, a few chapters will be dedicated to when and how to best utilize analogies in the psychotherapy process. Using analogies in the assessment process will be discussed, as will which clients will best benefit from certain analogies. A case study using a core analogy, encompassing a central dynamic with multiple therapeutic uses, is presented demonstrating how the use of a single analogy can be instrumental throughout the client's psychotherapy.

Chapter 1

The Advantages of Using Analogies in Psychotherapy

As previously stated in the introduction chapter, the essential advantage of using analogies in psychotherapy is the characteristic of being a *transitional language* between therapist and client. It provides the therapist with a mode of changing the therapeutic dialogue from one that is primarily a complex theoretically driven interchange to a more commonly understood language. This transformation in communication style provides several advantages.

Analogies transform the personal and vulnerable
dialogue to one that is impersonal and safe

In the initial rapport building stages of psychotherapy, the client is frequently unsure of the therapeutic process, including trusting how the therapist will handle personal material. Standard therapeutic language may cause the patient to feel vulnerable and prone to unnecessary resistance. The use of an analogy early in therapy to communicate vulnerable material allows the interchange to focus on the content of the analogy rather than the clients well-defended intrapsychic dynamics. This allows the communication to remain safe and impersonal until greater rapport is established.

For example if a client comes to therapy complaining of frequent unsatisfactory and transient relationships, the clinician may quickly gauge from a social history that the origin of these failed relationships is rooted in early sexual trauma or emotional neglect. In this common scenario, it is easy for the therapist to surmise that the client's behavior is at least in part resulting from an early fusion of acceptance and sexuality. The failure of these relationships has, at its core, a faulty mechanism for which to maintain self-esteem. In essence, the client fails to believe that she can be loved or accepted without the sexual component, thus allowing her self-esteem to be too dependent on the success of her relationships with men. In addition, the client may present in session as dependent and sexually provocative while merely hinting that she becomes prematurely physically involved in her relationships.

With only veiled hints of early sexual involvement in relationships one can imagine the vulnerability the client would feel if this topic, along with her in-session nonverbal behavior, was immediately interpreted. This is especially so if the material the patient offers in session focuses predominantly on the negative behaviors of her various suitors. To avoid early direct discussion regarding the vulnerable aspects of the patient's sexual promiscuity and this behavior's contribution to the relational failures, the analogy of the slow leaking ball and the air pump would be used to deal with the core issue while rapport is established.

In this analogy, the ball with the leak is compared to the individual with low self-esteem and the air pump is analogous to the mechanism in which one regulates self-esteem. It is presented in analogical form in hopes that the patient will come to see the leak as similar to the depletion of self-esteem felt when her partner withdraws affection. Hopefully she will realize that the hole is analogous to the original narcissistic injury and confusion of the earlier sexual trauma. Furthermore, by allowing men to be the sole regulator of her self-esteem (in this case the air pump) she relinquishes her own ability to regulate her sense of worth. The use of this analogy provides a starting point where the client and therapist can discuss personal and vulnerable dynamics by focusing on the impersonal and safe content inherent in the analogy. It also aids the client in identifying means to self-regulate self-esteem, to be their own self-esteem pump, instead of depending solely on attention from men.

Analogies provide an objective perspective until a greater depth of interpretation can take place.

Frequently the emotional origin of the presenting conflict in psychotherapy is simply too overwhelming for the client to deal with in the early stages of treatment. Analogies allow for the maintenance of an objective perspective until a deeper level of interpretation can take place.

In another scenario, a client enters therapy complaining of intermittent and explosive episodes of rage and anger. During the intake sessions the clinician discovers that the individual was subjected to frequent physical abuse as a child. There is a distinct correlation between the early physical trauma and the patient's present manifestation of rage and explosiveness. It also becomes clear early in therapy that the patient's defense against this terrible insult is to present on the surface an air of imperturbability and macho bravado. This individual is the definition of over-controlled hostility. Simply interpreting his behavior and stoic presentation as a defense protecting a hurt little boy during the early phase of therapy, the therapist runs the risk of the patient experiencing agitated indifference or early termination. Worse yet, regression with no real set of adaptive defense mechanisms to replace his existing ones, leaves the client vulnerable and unnecessarily exposed. It would seem imperative in this case that the client is able to maintain a cognitive or objective perspective regarding his core dilemma until he is more comfortable talking about his early abuse. In this situation the analogy of the shaken pop bottle might aid the client in maintaining a cognitive distance from the extreme emotional material while at the same time allowing him to work on therapeutic goals of reducing the frequency and intensity of his rage.

This analogy compares the individual's ability to express anger to the contents of a soda bottle under varying degrees of pressure. In session, the therapist may relay that individuals dealing with anger issues often resemble an unopened carbonated soda bottle. Each shake of the bottle, no matter how slight, causes the pressure of the contents to increase. If there is not a corresponding release to each increase in pressure, there is an accumulation that can explode after a time, even with just a slight lifting of the lid. Complicating matters, it seems impossible to tell from looking just how much the unopened bottle has been disturbed. The liquid appears still, regardless of the amount of times it has been shaken.

As one can see, the analogy contains many of the dynamics a sufferer of intermittent explosive episodes deals with. The accumulation of significant as well as everyday stressors is seen as analogous shakes of the bottle. The inability to possess more subtle and appropriate ways to express mediated doses of anger is in this case similar to not releasing pressure through periodically lifting the lid. In addition, the false appearance of the undisturbed liquid is synonymous to the imperturbable defense these individuals often present. Moreover, the opening of the oft-shaken bottle leading to the volcanic eruption of your favorite carbonated beverage is analogous to the violent display of rage after sometimes only the slightest provocation.

Using this analogy, treatment strategies begin to emerge that are rationally discussed and implemented while simultaneously building rapport where later discussion and interpretations of early trauma can take place. Treatment goals can focus on helping the individual recognize and keep inventory of the various stressors or 'shakes of the bottle'. Once the stressors are identified, the therapist can encourage the client to develop methods of periodically relieving pressure such as exercising, practicing a hobby, relaxation. Treatment goals can also focus on the training of mediated methods of expressing anger.

For example, the use of assertiveness training works well for those situations when the 'pressurized' bottle wants to explode. Perhaps only when the patient states he has reached a plateau in his therapeutic gains as one of my patients recently did, that therapy would be ripe for deeper discussion of historical material. "Dr. Seaton. No matter how much I try and relieve this pressure, it seems I'm always under pressure." It was at this juncture that I was able to venture a historical interpretation regarding the connection between his early childhood experiences and present manifestations of rage.

Analogies Allow the Client to Know
Where They Are Vis-à-vis The Therapeutic Process

Because analogies often have a beginning, middle, and end, or at least one that is implied, it is advantageous to the client to understand exactly where they are along the path to behavior change. In addition, as seen in the previous section, it allows the patient to gauge whether or not they are reaching their therapeutic goals. In this sense, it empowers the client to take control of his or her own therapeutic process and discourages unnecessary dependence. At any given point the client can gauge and report where they are along their therapeutic course, which allows them to take appropriate action up to and including termination of therapy.

For example when a patient enters therapy after struggling with a life changing event, they are cumulatively beaten down, confused, and without hope. Numerous negative consequences have resulted from this single transforming event and the patient is unsure of strategies they can incorporate to gain control of their lives. They want help, but are unsure of how their situation got so bad, much less how they are going to change it. In these cases I offer this analogy to gauge where the patient is in this ongoing dilemma, and what it might look like when they have reached their desired goal.

In this analogy the life altering event is compared to a leisurely walk along the edge of a river. Unexpectedly, they slip, and are unceremoniously dumped into the river's raging rapids. As they initially struggle with this unexpected peril, they do all they can to simply keep their head above the vicious force of the rapids. At this time, they cannot worry about the consequences of their falling in as they are swiftly drifting further and further away from their original destination. They must do anything they can just to survive. Adrenaline kicks in and they navigate the best they can, struggling for any means to escape. Just before it appears they will completely submerge, they make one last grasp for an exposed root from a tree at the river' edge. The grab is successful and immediately their trip down river comes to a halt.

Still not completely out of peril, the relief of establishing contact with mother earth is short lived, as the waters force continues to threaten to dislodge their grip. Eventually gaining enough strength, the individual is able to laboriously pull him or herself to shore. The initial experience of terra firma and deliverance from danger is exhilarating and the individual takes time to bask in their achievement. However, almost too soon after their self-rescue they come to the realization that in the midst of their harrowing journey, they have been swept miles away from where they began, and now must begin the arduous trek back to their original starting point.

At some time during their sojourn back to square one, a multitude of feelings begin to emerge regarding their ordeal. They are often angry about the circumstances that placed them in this situation in the first place. They sometimes feel hopeless about even getting back to the original starting point, and if they do, will things ever be the same as before? What seemingly irreparable damage has occurred during this event? Individuals entering your therapeutic setting having undergone a life-transforming event will experience all of these feelings and more.

In this analogy, there is an implied ending as most people eventually make it back to the original starting place. While permanently changed, they usually recover to lead productive lives. So when utilizing this analogy in psychotherapy, one can aid the client in gauging where they are along this process. Have they just been thrown in the river? What are the steps to get out of the raging aquatic force? Are they at the point where they are assessing damages before beginning their journey back to the original starting point? Have they painstakingly returned to their original place and are now wondering how to move on with their changed lives? No matter where they are along this continuum, there is an empowering quality to this analogy as the patient can see a possible end to this journey and visualize the next step. This analogy is particularly helpful for individuals who have experienced such life-transforming events as: a divorce after a longstanding marriage, a failed business, or an unexpected death of a child or a spouse.

Chapter 2

The Case of Teri L.

A young cub, one of a litter of three, was born to a lion pride on the African Sahara. This cub was much smaller than his siblings. Scrawny by comparison, he remained the runt of the litter throughout his young life. Although the runt, he had an uncanny ability to survive in these harsh circumstances, doing just enough to gain the minimal sustenance required to remain with the pack. His siblings, continuing to gain in strength and size constantly make him the mock victim in their instinctual play, knocking his legs from underneath him and simulating the strangling chokehold they will later use during the hunt. The cub perseveres, but it is quickly becoming apparent that this lion is developmentally impaired.

Then one day the cub, innocently exploring his terrain, finds himself separated from the pack. Despite the continuous calls from his mother, he does not hear, and as is the harsh survivalist way of the pride, they have no choice but to move on.

The scientists documenting the pride state the cub has little chance of surviving in this harsh environment, and will most likely be killed by rival predators.

Somehow, after almost two weeks, the emaciated cub miraculously reunites with the pride. What should be a joyful reunion is tempered by the now exponential gap in size between himself and his siblings. They are becoming adolescents, displaying the small tufts of back hair that will later become the full-fledged mane of the adult male lion. The pride has also expanded during this time, adding hungry new cubs dependent on the pride for their existence. Soon after, the siblings are ousted, earmarked for their own pride, or perhaps a later triumphant return to supplant those who have ousted them. The pride seems almost confused by this anomaly of the cub's survival. Against their natural path of instinct, the cub now enjoys a unique place. He is free to fraternize among the male leaders. Not sure what to make of this larger yet immature version, they continue to treat him as a cub instead of a potential rival. This expanded adolescence lasts during the rainy season and on into the dry months.

This languorous existence enjoyed by the cub is shattered in an instant when he attempts to take his usual feeding position at a recent kill. As he begins to feed along side the dominant male, the same male that he was playfully jostling with only days before, the larger lion turns his head in menacing fashion bearing his large canines and roaring ferociously. The startled cub leaps back three feet before slowly returning to the quickly disappearing carcass. This time the older lion attacks the cub leaving vicious claw marks, leaving no doubt of his intent. The cub is no longer welcome at the kill, no longer allowed to partake in a hunt in which he was not a participant. The confused and injured cub retreats to the females in the pride for support and comfort.

The females, following the lead of the dominant male, repeatedly rebuff the now adolescent cub's attempts for solace. Within a five-minute time span the young lion that had survived so much has been transformed from an overgrown cub, the pride mascot, to a potential threat to the prides very existence. He is now banished from the pride forever.

As the viewer watches the still thin and undersized lion wander off into the darkness, the scientists documenting these events state a grim prognosis for this supposed king of the jungle. As a rogue lion he was perhaps destined to a life on the periphery subsisting on carrion or small prey if he was to survive at all. The nature show ends with a cryptic statement from the narrator stating that in fact they never knew for sure if they ever saw the lion again.

When viewing the program, especially as the lion disappeared into the dark horizon, a multitude of ambiguous feelings struck me. A feeling of over protectiveness came over me. He was not ready yet to hunt for himself. Anger also emerged toward the pride. Didn't they realize what he had been through? Why hadn't the pride trained him better, making him part of the hunt more often? I could not relate to the harsh treatment they gave him at the kill. Didn't they feel guilt as they expelled him to who knows what fate?

The narrator chided my anthropomorphist thoughts and feelings, explaining that had the cub been allowed to stay, the delicate dynamics of the pride would be significantly altered, leading to situations that could be detrimental to its very existence. He stated that during the season of drought there was simply not enough food to ensure the survival of the pride, including the new cubs. In addition, when the inevitable coup took place with a new set of dominant males ousting the old guard, the cub would be banished anyway, perhaps killed by the reigning victors. Even if he did live, at this point his skills would be even less honed. A Darwinian pall washed over me as I struggled with my human feelings and obligations. Lacking the workings of the frontal cortex, animal's decisions become instinctually correct and simple in their execution. Our choices regarding our offspring are an inherently more difficult dilemma.

This was the case for Teri L., a mother of three, who I relayed this story to during the intake session of her psychotherapy.

Teri L. was a 46 year old successful banker and mother of three males ages 24, 21, and 17. Her youthful appearance belied the fact that she was a grandmother of a 2 year-old girl. She and her husband had just celebrated their 25th wedding anniversary. She arrived at my office for her first appointment seeking individual and possibly "family" psychotherapy.

During the initial session she detailed a family history that could be used as a textbook example in any course on codependence and substance abuse. Teri detailed her 24 yr old son's chronic history of polysubstance abuse that saw the family pay for numerous inpatient and outpatient treatments. There were reports of sporadic sobriety and numerous relapses with the identified patient being the impetus for frequent family dissention. Complicating matters was the out of wedlock birth of his daughter to a woman who he met in rehab who also vacillated between sobriety and abuse. Soon after the birth, the parents quickly demonstrated difficulty in caring for themselves and their infant. Teri and her husband agreed to take the three in on a temporary basis.

The addition of the new family brought added chaos to an already tenuous situation. The new couple fought constantly with the young mother often taking off for days at a time. Teri described all sorts of enabling behaviors not the least of which included taking over the role of primary care taker of her granddaughter. With this ever-increasing role came the inevitable joy and bonding experiences only a grandchild can elicit.

Months turned into a year. The middle son, a high achieving student (of course) became fed up and left for college. The youngest child, somewhat neglected and for years allowed to fall through the gaps, was failing in school and struggling with his own burgeoning marijuana abuse. This latter issue was one Teri's greatest concerns upon entering counseling, but it was the looming probability of the loss of her granddaughter's relationship that was the true impetus for seeking therapy.

The grandchild's mother had been clean for some time and maturely decided that the current environment (the father was still using on and off) was not the best for her child. She had secured a place to live with supportive relatives who had previous experience with the successful Alanon programs. Teri was devastated but with her son still abusing and no legally binding custody arrangement, she had little recourse in the child's removal from her home. She pleaded with her son to again enter treatment and researched prestigious recovery programs, even agreeing to pay once again for their costs. These were the presenting complaints that Teri brought to her initial psychotherapy sessions.

Teri's initial hopes were that I would be able to convince the eldest son for the last time to become sober, a task that therapists much more experienced than I had failed at repeatedly. She asked for advice on her decision to pay for her son's substance abuse treatment, and stubbornly held on to hope when told that the likelihood of this treatment success was limited. Despite my opinion, she and her husband once more paid for their son to enter rehab. During the third session of therapy Teri relayed to me that her oldest son did indeed sign himself out of the in-patient substance abuse program after only 3 days. Broke and out of state, he desperately begged Teri for enough money to return home, vowing again that this time he would stay clean if given another chance. As was her pattern, she reluctantly capitulated.

She also wanted me to talk to her youngest child, the struggling 17 year old, in hopes that speaking to "someone other than his parents" would reinvigorate his zest for academics. Like many patients entering therapy, the hope is that by changing the components of the system the internal dynamics will transform and equilibrium will be restored.

It became quickly apparent that taking this approach would be futile (I later did, at Teri's request, talk to the 17 year old. The eldest son scheduled an appointment but did not show.) However, Teri was at her wits end and needed some sort of framework. It was during the latter part of the first session that I offered the analogy of the lion cub and the pride to aid Teri in switching her perspective from that of a set of individual problems to that of a family systems approach.

I initially empathized with her dilemma comparing her decision to that of the lion pride. How does a family decide when to let go of a non-productive adult family member, especially one that seems so ill-equipped, handicapped by a mental illness, and unprepared to make it in the real world? This interpretation jolted Teri in that she had not really thought about making the decision to have her son move out. This was despite the fact that she realized that he had been an excessive burden, causing her to neglect other aspects of her family (i.e. her marriage, her middle son leaving, and her youngest son's developing problems).

I explained to her just as the nature program's narrator had discussed that allowing the lion cub (her oldest son) to stay would jeopardize the natural progression of family dynamics and allow her son to grow increasingly dependent, eroding the minimal functioning skills he already possessed. It did not make it any easier, of course, considering his present condition. Just as the lion cub would probably subsist on carrion, and maybe not at all, Teri's son would no doubt initially at best live a minimal existence. However, as most standard therapeutic approaches to chemical dependency suggest, the abuser often has to bottom out before actively seeking therapy.

Although Teri had heard the analytical jargon regarding codependency and substance abuse many times before, the analogy gave her a fresh if grim perspective on the central factors of her dilemma. It gave her a framework from which to structure therapeutic goals and measure progress. We would return to aspects of this analogy repeatedly throughout the course of psychotherapy.

For example, in discussing the harsh reality of the cub's banishment from the pride, Teri lightheartedly joked that when angry at her son she fantasized about executing the unexpectedly quick and precise threatening gestures of the dominant male. Instead, being the frontal cortex creatures that we are, we developed a structured exit plan that would incrementally increase independent living skills including, paying rent, and creating a budget and savings plan with a three-month period goal of moving out.

During therapy, Teri had to deal with her own harsh realities of how this situation had got to this point. Her own guilt regarding a previous depressive episode when her son was younger as well as the attachment to her granddaughter were identified as triggers for her continued enabling of her son. She had to visit the historical faults of her own marriage, recognizing that at least part of her codependent behaviors were mechanisms enabling her to ignore her ambivalent feelings regarding her husband. Part of that ambivalence was her husband's reluctance to be more authoritative with his sons and his own secretive weekend pot smoking. The utilization of conjoint sessions attempted to elicit support from this 'dominant' male in implementing this exit plan.

As therapy progressed Teri and her husband presented the exit plan to their 24 year-old son. Explaining that even though this action seemed harsh, they had made the decision to bet on him to make it instead of the implied belief that he could not succeed passed on by their previous codependent decisions. They did their best to explain how this decision benefited both him and the family. As predicted in a previous session, and again using the lion pride analogy, Teri's son, motivated by insecurity and fear, vehemently objected to the plan. Using all matters of previously successful manipulative behaviors, Teri's son, much like the soon-to-be banished cub seeking solace from the pride's females, hoped to maintain his comfortable status quo. Seemingly harder than it was for the lionesses, Teri struggled to rebuff these ploys and held fast to her difficult decision.

As the three-month period grew near, Teri found that,

despite her fear for her son's survival, she grew more confident in her decision. With her decrease in engaging her son in heated debates regarding his behavior and corresponding enabling behaviors, Teri found more energy and time to spend on other matters. She researched grandparent's rights to ensure an ongoing relationship with her granddaughter, focused on rekindling her marital relationship, and was able to implement consequential behavior with her youngest son. As the three-month deadline approached, there was definite improvement in Teri's emotional well-being and she expressed a hopeful outlook.

The weekend prior to the deadline for his moving out, Teri's oldest son came home extremely intoxicated and in a particularly foul and abusive mood. He chided his parents for their failures hoping to elicit maximum guilt. He stated that he was still $200 dollars short towards the apartment he had hoped to rent, and even threatened suicide. Teri called in crisis wondering what to do. We discussed her dilemma for fifteen minutes. She agreed to stick to her decision, recognizing her son's behavior from other previous arguments. Teri related in her next meeting that after telling her son her decision, he stormed out of the house.

Again, similar to the analogy of the banished cub, the ongoing fate of Teri's son is unknown. Six weeks after angrily bolting from the house, Teri was unsure of his permanent residence. She heard from her granddaughter's mother that he had been staying with a couple of friends and was demonstrating sporadic and inconsistent employment.

Three weeks later Teri terminated psychotherapy for the time being. Despite her continued anxiety regarding her son's fate, she recognized that it was at least subsiding. Not only was she surviving, she reported continued improvement in her family's functioning. Her youngest son was improving his second semester grades, and her granddaughter's mother, still sober and supportive of Teri's decision was facilitating a relationship between her daughter and Teri.

One can see from this case study the utility of this particular analogy in helping the client reframe their situation and structure treatment goals. Teri of course told me this during the termination session, and thus its inclusion in the book. It may be helpful to go back and reread the analogy, evaluating how one might use this analogy differently had therapy taken a different course.

101 Useful Analogies

Without further ado, here are 101 therapy-tested analogies to provide a fresh perspective for your clients as they undertake the valuable endeavor of psychotherapy. These analogies are a transitional language that should aid your patients in reframing their clinical issues, providing an objective holding ground until deeper emotional material can be explored. It will also provide a structure that includes a proposed beginning middle and end to their problematic situations. Some will be used simply for emphasis; others to aid the client in identifying where they are in the therapeutic process. Many of the analogies have various offshoots and can be referred to repeatedly throughout the therapy process providing specific treatment goals and providing the client a structured way of identifying where they are on the way to achieving them. Categorized by problem area (marital, parenting etc.) these analogies come with highlighted text boxes that demonstrate what constructs are relevant and which clinical situations are most appropriate.

Marital/Relationship Analogies

"Pull me up from the cliff's edge and I'll promise you anything."

- Marital betrayers' desperate pleas to do anything to return to the relationship are compared to the desperate pleas of an individual hanging from a cliff.

- For individuals needing to set boundaries before letting a transgressor return to the relationship (i.e. extra-marital affair, substance abuse).

- Boundary setting

- Abandonment anxiety

This analogy benefits those individuals in a marriage or relationship where a pattern of significant betrayal occurs (i.e. substance abuse relapse, physical abuse, extra-marital affairs) and the client has initially asked his or her spouse to leave. As time passes following the behavioral occurrence, the offender asks for forgiveness and another chance, promising to do whatever is required to get back in good graces with his partner.

The victim of this behavior, the immediacy of the emotional injury diminishing over time, often finds himself or herself in the dilemma of wanting to go against their better judgment and allow their offender to return.

The victim, intellectually at least, often knows what is required to improve the chances that the offending behavior will not occur (i.e. substance treatment, anger management etc.). Of course, the offending partner knows too, often promising to start these endeavors as soon as they are allowed back. However, as human nature would have it, desperation or abandonment anxiety often motivates this pleading. The urgency of these promises frequently diminishes once the relationship is back to its old equilibrium. The following analogy often aids the client in dealing with the emotional pleas of the offender and helps them stick to their original and most often intuitively correct decision.

If I were dangling perilously over the edge of a cliff, hanging by only my fingertips, I would promise you anything if you would just pull me up to safety. If I was a contractor, I would promise to build that addition to your home that you had been unable to afford. Under the present circumstances, I would genuinely intend to fulfill my promise. Once you pulled me up, I would be so grateful for my return to terra firma that I would agree to start construction that weekend. For the first couple of weekends, I would arrive early working diligently until dusk.

However, after a time, increasingly removed from the emotional immediacy of my prior danger, I would slowly realize how much these weekends were taking me away from my own life, my own relaxation. I might begin to cancel, giving rational and reasonable excuses as to why I could not come that weekend. Eventually my initial flood of devotion to the project would diminish to a small trickle, the renovation left in a stage of partial construction.

The general thrust of this analogy is to aid the client in seeing that offender's emotional pleas, while at the time probably genuine in intention, are motivated by desperation and likely to diminish once the emotional urgency subsides. This particular analogy demonstrates the need for empowering boundary setting. In practical use the patient may become impervious to the offenders strong emotional pleas. Unfettered by emotional guilt, the patient may require their spouse to first undertake an appropriate strategy before they are metaphorically "pulled up". For example, a patient may necessitate their spouse or partner not only schedule an appointment for a substance abuse evaluation (can't pull them up too early) but require program completion as well as maintaining six-months sobriety before they are allowed back in the home.

"A log jam can be relieved by removing only a couple logs."

- Compares a couple's initial impression of marital counseling to an aerial view of a log jam, demonstrating the necessity to take into account that the flow of water underneath requires only a few logs/issues be removed to restore the natural flow of the relationship.

- Aiding couples in reframing marital counseling as manageable

- Solution-based strategies

An aerial view of a river's logjam often depicts an overwhelming array of twisted and impossibly stacked timber. From this vantage point, the task of relieving this logjam seems daunting. What one fails to recognize from this angle is the enormous force that exists just below the surface.

This force, constant and always moving forward, is the steady and continuous flow of the river. By using knowledge of a river's dynamics, one realizes there is no need to reposition each individual log. One must move only a couple of strategically important logs until the natural force of the water takes hold, restoring the river's unabated flow. Couples often enter therapy with what appears to be a logjam of a relationship. By the time they realize the need for marital counseling, this impasse has been severely impacted with each and every negative event stacked against each other until it must seem to them much like an aerial view of a logjam. They come into session thinking that every negative event, every oversight, every undesirable aspect of the others personality flaws must be dissected in order for the marriage to get back on track. Thinking that they must reposition every log significantly affects motivation to complete such a seemingly overwhelming and lengthy task.

By offering the couple this particular analogy, they can reframe this task recognizing the need to work through only a few central dynamics in order to break the impasse. In this case, the force of the river is analogous to the natural positive force of the partnership that will carry a way the smaller problems or debris along the way.

The important strategically placed logs are seen as the central dysfunctional dynamics that will be focused on in therapy. The therapist in this analogy is akin to the expert in water dynamics who will aid in choosing which strategic problems to focus on such as communication difficulties or family of origin issues. This analogy also has utility in individual therapy as well for those individuals who are overwhelmed by a particular event or decision. Getting them to prioritize which strategic logs they have to reposition diminishes the overwhelming anxiety and allows them to see the situation as manageable.

"When playing with matches, it's not good to be sitting in a puddle of gasoline!"

- Sitting in puddles of gasoline flicking matches at one another is akin to arguing about seemingly innocuous details while feeling unconscious and unresolved hostility.

- Utilized with couples to stress the importance of communication beneath the surface or conscious level. Used to demonstrate how unconscious feelings fuel surface or content level communication.

- Unresolved hostility

- Communication: content vs. latent intent of message

Couples often seem perplexed as to how mundane disagreements such as taking out the trash can escalate into explosive and emotionally abusive arguments. As therapists we know that verbal interchanges between individuals are not always about the overt content of the message, but instead express an underlying implication of unresolved hostility. Couples need to understand that a majority of communication takes place below the simple content level. The following analogy demonstrates this dynamic.

If we were sitting around flicking matches at each other, occasionally one might land and inflict a slight sense of discomfort before we could knock it off. However, if we were all sitting in a puddle of gasoline, that same match would likely ignite, causing significant damage.

Using this analogy during the course of therapy helps the couple reframe their situation, identifying central issues of unresolved hostility they feel towards the other. It also demonstrates a realization that during even simple arguments these feelings are never far from the surface. These puddles of gasoline, although subconscious in nature, can often ignite with only the slightest provocation (the simple disagreement, or match in this case).

The couple's recognition of the presence of these underlying dynamics is paramount in ameliorating long standing dysfunctional communicating patterns. Understanding this analogy enables the couple to mediate the intensity of their between-session arguments, determining if the normal day-to-day miscommunications are in actuality igniting underlying hostility.

This recognition of the fuel source (and its removal) can often prevent the simple argument from exploding into an hour-long debate. More than a few couples have reported that simply repeating the analogy during an escalating argument serves to extinguish the intensity.

"The Gift of the Magi"

- The need for unconditional giving in a relationship is akin to the classical Christmas story where a couple sells their prized possessions to give their spouse an unprecedented gift.

- This analogy is used when couples reach an impasse where they are both waiting for the other to act first before responding in a positive way.

The deterioration of trust in a long-term relationship leaves both partners suspicious of believing their significant others intentions. Throughout the course of any relationship there are a number of minimal to moderate or even serious betrayals and transgressions that leave the couple worn out and fearful of being the first to give in. Instead, each individual waits for the other to change their behavior, constantly disappointed in the other person for consistently failing them.

As a therapist, I of course empathize with the realistic concern of either partner further betraying the other. After validating their feelings, I stress that despite this reasonable concern, the strategy of waiting for the other to give in first will continue to result in perpetual disappointment and hostility. While it may seem overly romantic, I offer a new strategy in getting them to start at stage one of rebuilding trust by retelling the Holiday classic "The Gift of the Magi".

As one recalls the idealistic yet financially strapped couple both possess items of deep sentimental value: he an antique and expensive watch, she beautiful flowing locks. He desires a gold chain, she a comb and brush set. As most of us remember she cuts her hair, selling it to purchase the watch chain, while he sells the watch only to buy the brush and comb.

Although simplistic and idealized, it is a touching example of selfless devotion to another. Sometimes our most jaded couples require a little elementary positivism. An example of unconditional giving is a welcome change to the standard, non-trusting, and hoarding dynamics typical of these interactions.

"Relationships are like tributaries leading to the sea."

- Compares tributaries and their unabated path to the sea to the natural progression of romantic relationships in that they are constantly evolving and have distinct points of starting, and ending.

- Best utilized with couples at a relational impasse.

- Used when couples are frustrated at their attempts to return the relationship to a previous phase of development.

- Effective in dealing with fear of commitment

Commitment issues are often a central focus of unmarried couples. Invariably one partner is further along the commitment continuum, expressing anguish toward the lagging counterpart. Many couples struggle with the inevitable flow or changing course of their relationship. Therapists often hear the complaint, "I want our relationship to be the way it used to be, when we had fun." Another complaint involves one of the couple dragging their feet. Usually these commitment fears have, at their core, unresolved childhood dynamics or previous breaches of relational trust. Highly developed and ingrained defenses

41

outside the individual's conscious awareness guard these fears against any possible intrusion. This often leads to intellectual rationalizations concerning why one cannot commit just yet. The couple enters therapy confused by this relational impasse, unsure of what direction to take. The following metaphor provides a framework for the natural course a relationship takes, complete with the multiple options each individual possesses.

Relationships are similar to tributaries leading to the sea. There is an inevitable flow operating independently, regardless of what each individual brings to the relationship. A relationship, like a tributary, also has a preordained destination. Just as a tributary empties into the sea, a relationship's end point is lifelong commitment. Each participant, upon entering the relationship flow, whether they know or it not, is bound by the metaphorical force that is analogous to the physical laws that govern a tributary's course to the sea. At any time either or both individuals can simply exit the flow. They can also to choose to negotiate the various rapids and river bends, gaining force and knowledge along the course to their inevitable destination.

Usually couples find themselves at an impasse when they are choosing an option that goes against the relationship's inevitable flow. Often one of the partners treads water attempting to remain in place. At first, this technique proves simple and effective against the subtle down stream flow. After a time the constant effort becomes tiring, and the partner not fighting the current becomes increasingly aware of the ever-widening distance in the relationship.

In treatment it is helpful to aid the client in identifying that he is indeed treading water, hopefully helping them in gaining insight into the possible reasons for stalling in the relationship. Many times it is both individuals who are treading water, afraid to negotiate the unknown of further commitment, but not wanting to face the abandonment consequences of exiting the river. Sometimes in these cases both individuals are denying they are treading water, going about business as usual hoping something will unexpectedly come along to boost the relationship.

Couples attempt other strategies as well to ward off relational impasse. One or both individuals may attempt to swim upstream trying valiantly to recapture the magical feelings only the beginning of a relationship can elicit. More often than not, these attempts are thwarted and the couple, unable to adequately swim upstream, is frustrated as the water's force returns them to their original place along the journey. Complaining that their relationship does not have the same spark as it used to, they are simply dealing with new developments and expectations the relationship has picked up along the way. For example, a previously divorced couple may begin a torrid, on-the-rebound romance that provides weeks of uninterrupted satisfaction. It is only when they begin to incorporate their children into the relationship that it hits a snag. Unconsciously noticing that this intermingling portends future conflict, the couple instead denies this possibility, attempting to continue the status quo.

This changing development prevents duplicating the previous nature of the relationship. No matter how hard a couple tries to swim upstream they will inevitably end up at a place where they will have to negotiate this curve in the river. Again, it is the therapist's job to help the couple recognize their dilemma and hopefully aid them in continuing.

"A Jaguar? Sure they take a lot of maintenance, but when they're working, there is no better ride."

- The maintenance of a destructive and fatalistic relationship is analogous to maintaining an exotic automobile vs. a more practical model.

- Best utilized for individuals involved in a fatalistic relationship with an individual possessing Axis II characterological traits.

- Object Relations Theory: part–object relation

- Idealized fantasy relationship

Two or three times a year an average guy comes into my office devastated by a relational predicament involving a classic femme fatale. This ordinary male, sometimes married sometimes not, has unfortunately become entangled with an exotic and mercurial female. Intoxicated by something he has no business possessing, he is unable to resist the allure of this exotic temptress, blind to obvious flaws and suspicious surrounding circumstances.

Upon hearing the description of the various young ladies, the objective clinician finds it easy to quickly form a diagnostic impression by proxy involving clearly characterological origins. Nonetheless, I am constantly fascinated by these indeed exciting and interesting women, clearly understanding these gentlemen's intense infatuations. Some have even brought in pictures, and I am no longer shocked to find that these women are exactly as beautiful and exotic as my clients have described them.

Usually the patient reports the most enjoyable and exciting time of his life, followed by an unexpected yet significant cooling of the relationship initiated by the female. During the next couple of months, the male finds himself scrambling, exerting whatever degree of emotional or financial energy needed to restore that previously unparalleled existence. Characteristic of the borderline personality, these relationships are usually intense, transitory and manipulative. Therefore, despite his efforts, the relationship never regains its original luster. Sure, there may be occasional glimpses of the initial intensity, but usually the men find themselves emotionally and sometimes financially depleted.

These men enter treatment hoping to find a way to recapture the initial brilliance of the relationship or at least understand why these women operate the way they do. The ultimate goal of treatment is to eventually aid the client in gaining insight into the unconscious motivations of their initial attraction and maladaptive maintenance of this destructive relationship. First, the therapist must get the patient to recognize that any further investment into the relationship provides little return. To accomplish this, the therapist must first validate how tremendous this relationship was. Often I relay an early car buying experience to emphasize this understanding.

Upon leaving graduate school, (owing a sizable figure to some woman named Sallie Mae) I had approximately $6,000 in which to buy an automobile. Wanting to at least convey some sense of suave cachet to go along with my newly acquired PhD, I bought a car trader magazine and conducted a cursory skim past pages filled with Honda Accords and Nissan Sentras stopping abruptly at a page containing possibly the coolest ride in the universe. A professionally angled photograph of a 1978 British Racing Green Jaguar V12 XJS coupe all but pulled me into the picture.

Impulsively, I gathered my car mechanic friend and we were off to visit my future wheels. I ignored my friend's initial smirk as we approached the vehicle, paid only half attention to the seller's spiel, and hurriedly took my seat behind the British masterpiece. It had an awesome feel, lean and low to the ground. The test drive was just as satisfying and even my friend had to marvel at the powerful low growl of the engine. Tempted by the smooth clean lines of the exterior and the allure of the exotic, it was all I could do to not write the check while negotiating turns.

My friend was a little more than casually persistent that I at least take some time to think rationally about my decision. On the way home, my friend kept a cool head as I passionately extolled all the advantages of such a ride. He finally chastised my naiveté stating, "Sure a Jaguar is a cool car, and it would be great to drive a real status symbol, but do you have any idea how much maintenance it takes to operate one of these cars? They are notorious for frequently being in the shop, but I guess you could tell your fellow passengers on the bus what a great ride it is on those days you get to drive it."

What stood out in his statement was the hint of sarcasm in the way he said, 'status symbol'. I was immediately defensive and wanted to accuse him of feeling jealous, but there was too much truth in what he was saying. I chose a 1984, 4-cylinder Ford Mustang that was reliable transportation for the next few years as I made a dent in my school loans.

Years later I still try to figure out how to buy that Jaguar (the experience of that first test drive has never left me). However, my objective research always tempers my enthusiasm. As much as the pleasure seeking mechanisms want to convince otherwise, the rational executive functions of the ego remember that much of what my friend said proved to be correct.

"A tree that weathers the storm does not stand rigid but instead bends with the wind."

- Defensive rigidity, with fixated and limited coping skills is comparable to a tall and rigid tree's susceptibility to falling under an intense storm. Conversely, it compares trees able to bend in the wind to individuals with defensive flexibility.

- Best used in couples where one member possesses rigid and inflexible defensive mechanisms as a means of defending against inadequacies and insecurity.

- Defense mechanisms

Couples often possess rigid rules and expectations regarding how a relationship works. One or both parties have established these stringent doctrines, usually as defensive means to ward off the anxieties of dealing with the grey areas inherent in any relationship. Usually this is the product of a restricted repertoire of coping skills. One partner may attempt to impose these unbending expectations on the other. This dynamic is often marginally adaptive in situations involving low stress. However, unforeseen life-changing events often render this coping skill useless. Under these circumstances the individual may initially adapt by becoming even more rigid, further burdening the now fragile relationship. This individual in therapy is not likely to accept an initial interpretation that their rigidity is a product of limited coping skills. An intermediary analogy may be helpful in lessening the intrapersonal impact.

In a severe storm, the tall, rigid trees would seem on the surface to be impervious to the strong winds. On the other hand the smaller, thinner trees seem likely to succumb to even the smallest surge in winds. The fact is sometimes the most rigid trees cannot bend to adapt to the strong and changing winds and these massive trunks are easily uprooted. Surprisingly, because they can bend in all directions, the thinner trees can often survive even the most vicious storms.

"Reading the rings of the tree"

- This analogy compares transient relational crises to reading the crosscut rings of a tree.

- Useful in aiding couples to view their relationship with a longitudinal and historical quality.

Couples frequently enter therapy because they are currently experiencing the effects of a serious relational trauma, (infidelity, financial crisis, estrangement from family). They are so ensconced in the event, clouded by the exacerbating stressors, that they lose perspective of anything in their relationship that came before, or about the prospects of anything possibly occurring after the resolution of the current conflict.

It is important as a therapist, as a means of eliciting therapeutic motivation to aid clients in reframing this 'forest for the trees' perspective. In doing so, it is important to aid the couple in visualizing their relationship as having a longitudinal historical quality. To achieve this I often compare their relationship to the dissected trunk of a tree.

As we all remember from Junior High science, the various rings seen in the dissected tree trunk represent distinct periods or years of the tree's existence. The varying degrees of thickness signify whether it was a good year, or a year of drought. Invariably when looking at most tree histories one can

51

see good years followed by drought years followed by good years and so on.

The implied message of course is that most relationships will have troubled times, but it is important to remember that there were previous storms weathered and good years in between.

"The need for space during a separation is akin to mirroring a child."

- The necessity for a recently left spouse to allow contemplative "space" from his partner is comparable to the mirroring of a toddler experimenting with autonomy.

- This analogy is used in situations where a usually passive-dependent spouse has separated from the relationship after years of oppressive treatment. The desperate spouse changes their behavior, but it serves only to further distance the other.

- Passive-dependence

- Abandonment anxiety

- Self-psychology: mirroring

- Autonomous functioning

Patients often enter psychotherapy with this common scenario: a spouse after years of enduring emotional abuse decides to abruptly leave. This individual has usually demanded change in their spouse. These protests routinely ignored, they finally make good on their many previous threats to leave. The deserted spouse, suddenly panic stricken with abandonment anxiety, vows to change. Although the intent may indeed be genuine, no manner of demonstration of this new behavior makes much of a dent. The individual claims they simply require "space" to contemplate the status of the relationship. The individual persists in putting their best behavior forward to no avail, and in fact, their attempts only serve to 'smother' their spouse, causing them to distance themselves even further.

The chagrined and desperate client enters your office desperately needing you to help them get back their spouse. They consequently cringe when you suggest that they validate their partners need to have some space, telling you the risk of doing so imperils their relationship even more. In this situation I validate the risk, but I suggest that constantly hounding their spouse with good intentions smothers them even more, activating the distancing behavior.

They express confusion that occasionally during the separation the two had spent some quality time together and they had pleasingly observed the 'changed' behavior only to become more distant the next day. In most cases, the partner, usually with a history of being passively dependent, is afraid that if she lets her guard down there will be a return to status quo. Thus, she is likely to run away from continued harassment no matter how positive the behavior is. As a means to explain this, I offer this analogy: When a child first realizes his mobility, he is almost drunk on this new independence. They sometimes runaway from the parent and contemplate dangerous new maneuvers of independence. If the parent runs hurriedly and anxiously after them, the child often makes a game of this, running even further away. If the parent never lets the child experiment with this behavior, the child may develop a false sense of security never getting to experience the negotiation of failure or the occasional pain of unrestrained freedom. They may become frustrated later on at their parents constant smothering.

While indeed there is a risk that the spouse will take this space and actually seek a divorce, more often, they reevaluate their position and return after seeing sustained behavioral change in the other. That is why I encourage the desperate spouse to truly allow for ample space so that the partner asking for more space can accurately get a glimpse of what it would be like to be on their own.

Similarly when the child who has ventured off realizes that they have gone too far they turn to see where the parent is. If the parent mirrors back approvingly, the child will usually venture only a bit farther, aware of the distance. If the parent is right behind them, the child will only run ahead further. This analogy can help explain to the patient the erratic behavior of their spouse and allay the anxiety of their situation.

"In order to boil water sometimes you have to turn up the heat."

- Turning up the heat to make water boil is analogous to the need for the passive-dependent spouse to assertively set drastic limits and boundaries if they really want to see change in their spouse.

- Used in situations where the client consistently complains about an underperforming spouse, but due to fear of independence, never requires them to change.

- Passive -dependence

- Assertiveness

- Learned helplessness

A common situation in psychotherapy finds the patient seriously contemplating leaving or at least temporarily separating from her marital situation. These individuals have usually endured years of accumulated mild to moderate emotional abuse. They have half-heartedly required that their spouse change their behavior, making many lukewarm threats to leave if the changes do not occur. However, afraid of the unknown elements of independence, they rarely follow through. The other spouse, often meaning to change, becomes staid and comfortable in the notion that their spouse will never follow through. This results in further erosion of respect for their spouse and continued sub-par behavior towards them. These passive-dependent individuals constantly complain about their spouse, but rarely take responsibility in propagating this dynamic. As a means to interpret this pattern I sometimes compare it to watching water under a low flame expecting it to boil and then complaining when it does not. I suggest that one might have to turn up the heat in order to get the desired results. This interpretation often leads the client to identify those behaviors that would turn up the heat. In this case the heat is akin to those behaviors (which may include separation) that would light the fire under one's buttocks, so to speak, to provide impetus for real change.

"Mussolini's dangling and bloated corpse. If only things could have been different."

- This multi-faceted analogy compares reparation issues in an extra-marital affair to desecrating Mussolini's corpse.

- Uses include couples dealing with an extra-marital affair, or any situation requiring the reestablishment of trust after a significant betrayal.

- Narcissistic injury

Undoubtedly one of the most difficult marital situations in which a therapist has to intervene involves that most sacred breach of trust, the extra-marital affair. The initiation of marital therapy under such emotionally devastating circumstances proves to be quite a delicate venture indeed. There are so many factors and feelings simultaneously occurring that it is often difficult to find a starting place.

Once, while thumbing through my father's Time-Life series of books on WWII, a provocative and disturbing photograph caught my eye. In the grainy black and white picture two unrecognizable bloated corpses hung upside down, there feet and hands bound by rope tied to what seemed to be a lamppost. The caption identified the bodies as being Mussolini and his mistress. It went on to report that following his death, angry villagers unceremoniously dragged Mussolini's body into the town square desecrating it for days before it was finally dragged in by a relative.

Somehow, this image instantaneously came to me while listening to a couple detail emerging feelings and consequences resulting from the discovery of the husband's recent affair. He was complaining that his dalliance with the other woman had been over for weeks and he repeatedly expressed how sorry he was. Still, he protested that his wife continued to bring up the affair, asking questions whose answers were sure to increase conflict while serving only to lower her self-esteem. Relaying the story of the Time-Life photograph allowed me to address the husband's failure to recognize that simply ending the affair could not prevent long lasting, even hostile consequences. Simultaneously it allowed me to help the wife see that while her righteous anger justified an extraction of more than a pound of her husband's flesh, there would be a time when continued desecration of the corpse would be unfruitful. It would never bring back the previous sense of trust any more than beating the corpse would ever bring back those villagers' sons and fathers killed by the now deposed and rotting dictator.

I have expounded on this analogy further proposing a scenario where Mussolini might have changed behaviors midstream. This expansion allows me to aid the couple in realizing that any chance of repairing a marriage post-affair must tolerate a high level of mistrust for a circumscribed length of time and that both parties must be vigilant in their patience if they are to negotiate this tenuous existence. This would include the offending partner offering complete transparency to the other including, frequent phone calls inquiring about their whereabouts, access to phone logs, texts, and emails.

In the proposed scenario, lets suppose that before his death, Mussolini delivered an impassioned speech from his familiar balcony location. In the speech, he would emotionally express contrition, stating he had recognized the errors of his ways. He should have never formed an alliance with Hitler, never had those who opposed him taken away in the middle of the night and shot. He would go on to express how his love for Italy would now be above his selfish concerns and he would implore his constituents to trust him and join him in his new agenda.

For the sake of this analogy's therapeutic benefit, let's suppose the convictions expressed in his speech were genuine. One can only imagine the length of time it would take for the people of Italy to believe him. Any slight variance or perception that he was reverting to his old ways would be highly scrutinized and subject to emotional accusations.

For example, he might in an attempt to sever all relations with Nazi Germany, wire a known Nazi to close any open accounts. The hyper-vigilant Italians would erroneously construe this transaction as continued relations with the Third Reich.

This is analogous to the tenuous situation a couple finds themselves in regarding post-affair adjustment. If the perpetrator of the affair is 5-10 minutes late coming home the other person often becomes highly anxious triggering those emotions inherent in betrayal. This analogy has served countless times in helping the couple understand that rebuilding trust takes consistent patience and near unconditional disclosure.

"Only one sperm gets to fertilize the egg."

- Remaining in a destructive and maladaptive relationship is similar to a single sperm fertilizing the egg to the exclusion of all others.

- Best used in counseling clients hopelessly stuck in a maladaptive relationship.

- Psychodynamic: repetition compulsion

On occasion, we have had those patients who remain in the most self-destructive of relationships. Despite any number of interpretations given across theoretical orientations, these clients desperately cling to chronically dysfunctional situations. They enter the therapeutic process hoping the therapist can provide insight into the never-ending cycle of breaking it off with the other individual only to reconcile at some later point.

Frustrated as to why they cannot shake off this particular person, they complain that one reason they hold on to these maladaptive relationships is their inability to find any nice guys/girls'. Although oblivious to the client, the therapist (and probably most of the individual's friends) recognizes that there are plenty of eligible people in which to become involved. Unfortunately, the lingering toxic relationship quickly turns off any healthy suitors. The involved parties, so caught up in maintaining the destructive relationship, are simply unaware of the existence of 'good' men and women often right under their noses. In an attempt to aid the patient in accepting that their own inadequacies are essential to maintaining the failing relationship, I frequently present this informative, if somewhat provocative analogy.

The propensity to remain in unproductive relationships can often be explained by the psychodynamic concept of repetition compulsion. This seminal construct posits that many people remain in maladaptive relationships in order to retrospectively repair unresolved longings in earlier primary relationships. For example, a daughter abandoned by a rakish father may choose a present mate with similar wayward characteristics. In changing her current beau, there is an unconscious wish that she will be able to retrospectively repair her relationship with the primary object (in this case her father).

As we know, millions of sperm fight for the privilege of fertilizing the female egg. Once the egg accepts but one sperm, the cell wall becomes impenetrable to all others. I remember seeing films showing slides of the recently fertilized egg. They depicted the remaining sperm, their tales undulating furiously, as they unsuccessfully butted against the perimeter of the cell wall. After a while, the sperm simply die off one by one leaving the one egg and sperm to commence cell division.

I relay this analogy in hopes of getting the client to see that as long as they remain in this destructively protracted relationship the wall for potential healthy suitors is impermeable.

"If you just keep walking up the hill, they will need to walk twice as fast to catch up."

- This analogy compares waiting for a capable, yet lagging hiker, to enabling a lazy spouse.

- Codependence, Enabling

- Empowerment

- Reframing cognitive schemas

A forlorn female fills out the prerequisite information in preparation for their initial evaluation for couple's therapy. As time runs past the hour, they assure you their partner is coming and request we hold off for just a few more minutes before starting? Invariably it gets to be 15 minutes past the hour and their spouse never shows. They detail to you a sad state of affairs. You empathize with this cheerless woman sitting across from you; the transparency of her inadequate self-esteem is almost tangible. If I can convince this woman to continue therapy even in the absence of her husband I can often empower her to elevate her self-esteem. An analogy I offer demonstrates how improved self-esteem alone can motivate change in her marital situation.

In this analogy I relate a story about hiking with my then 7-year-old son. I would invariably find myself out pacing him, and he would literally sit down in the middle of the trail and begin to complain. I would prod and coax him knowing that he was not yet tired. I would get him going and then turn and commence hiking. After taking but a few strides I would turn around again to see him sitting down in the middle of the trail. I would back track again pleading with him to continue. He would simply complain, before maybe taking a few more steps. This stop and go continued and no amount of cajoling would get him going. Then I came upon an idea; instead of stopping each time he quit, I simply kept going. Of course, I felt a little worried that he would not continue and he would fall far behind, or simply quit. Much like the woman in therapy worried that her husband would not try or worse yet, let her go. However, much to my surprise, my son would come sprinting up the hill his little legs churning double time. Not only would he catch up, but he would also sustain a greater effort throughout the remainder of the hike.

This analogy implores women to reframe their seemingly tenuous situation by correcting a fundamentally inaccurate cognitive schema. These women often sustain an automatic and maladaptive thought pattern that essentially posits that they are unlovable and undeserving of equitable and respectful behavior. Too often, they bargain and barter for even minimal acceptable behavior from their spouse. They are afraid to expect more; frightened their husbands will choose to leave instead of changing.

Women, who choose to heed the advice in the analogy, find they possess significantly more leverage than previously believed. These ingrained schemas are resistant to change and women frequently report trepidation in imagining trekking up the hill hoping their husbands will follow. To assuage these fears I have them imagine a win-win situation. One situation sees them reaching the top without their spouses, a situation where they are still better off. The other situation entails their spouses running twice as fast to catch up to their newly empowered wife.

Do you think this generation of the Hatfield's and the McCoy's really know what they're fighting about?"

- This analogy compares the constant feud between the Hatfield's and the McCoy's to the often ingrained and hostile communication patterns of marital couples.

- Used in marital and couples counseling where couples frequently fight about unresolved transgressions perpetrated against each other.

- Active listening

This following scenario is all too common during an initial conjoint session: a couple walks in loaded for bear, and literally within seconds, they begin a long diatribe against the other armed with historical dates of multiple transgressions. The other party launches a counter-attack correcting the others inaccuracies about who started these events, complete with explanations for their behavior. These couples have a long history of inflicting emotional damage to one another. They are not so much seeking counseling from you, but are rather looking for a judge to hear their grievances and render some sort of settlement. It is usually apparent in the first session or two that to trace back the chicken or the egg scenario will be fruitless.

I often propose a truce at this point. I bring up an analogy that most are familiar with, asking the couple to suppose that the Hatfield's and McCoy's are now in their 5th generation of their infamous feud. I ask the couple to suppose that despite the tenacity with which these two families fight, do they think anyone knows the real genesis of their feud? Furthermore, even if they could recall the reasons, do you think it really matters at this point? I suggest that maybe that is all they know, and maybe they do not really want to know what started it all, because then they would not have a purpose. Alternatively, maybe they are taking anger out on each other that actually comes from another source. This does not always waylay the couple's feud entirely, but it often slows them down enough to consider their method of communication.

The application of this analogy usually comes in the form of an in-session intervention that attempts to refocus the couple's communication, which has diverted from more meaningful problem-solving approaches. For example, in session a couple might be negotiating the division of household responsibilities. The communication patterns are progressing smoothly, beginning to approach a workable solution, when one of the parties adds a passive-aggressive zinger, such as, "You know, like that one year when you barely got off the couch." The meaningful communication, now thrown off track, soon deteriorates into the maladaptive default positions of the past. Once they are familiar with the analogy, the mere mention of The Hatfield's and the McCoy's usually allows for quick redirection back to the healthier lines of communication.

"Panning for gold"

- The steady yet sometimes painstaking process of panning for gold is akin to attaining healthy measures of self-esteem.

- This analogy is used in situations where one needs to change seeking unhealthy means of self-esteem validation (i.e. extra-marital affairs, abuse of pornography, gambling) to those methods that are more psychologically adaptive (i.e. relations with same sex friends, hobbies, community involvement).

- Self-esteem regulation

- Preventing and understanding extra-marital affairs

Helping patients recover from an extra-marital affair, or dissuading those who are contemplating such a transgression, is often challenging work. Many times the individual who perpetrates such an act is not a serial offender or 'player', but instead is someone who has difficulty regulating self-esteem.

Because of early childhood relationships and subsequent object relationships, the individual relies too heavily on attention from the opposite sex for validation. Perhaps from being mother's little husband or daddy's darling favorite, these individuals have developed a propensity to seek out and to a degree unconsciously manipulate relationships in order to gain acceptance or validation.

During periods of marital stress they are often vulnerable to utilize, although out of their conscious awareness, those seemingly automatic behaviors in which to quickly regulate self-esteem. These situations develop in an often fantasy-like state, with the offending party unable to fathom far-reaching consequences. I have found that these individuals often lack in their repertoire other, while containing lesser valence, modes of attaining self-esteem. In the therapeutic setting, it is the clinician's job to educate the client in methods of healthier regulation of self-worth. This expanded repertoire of self-esteem attainment carries significantly less potential risk to their marriage.

After the client demonstrates significant insight into the genesis of their searching out validation from the opposite sex, I explain, using an analogy, the necessity of gaining healthier means of pursuing self-esteem. I then compare self-esteem attainment to types of mining for gold. One can either risk going into the depths of a mine to attain a sizable gold nugget (the electric buzz of attention from a new person of the opposite sex) or taking a pan, and in a less alluring but safer manner, working to collect the same amount of gold dust (real and lasting self-esteem).   Sustained panning yields equal value to the gold nugget.

For example, ways of panning for gold may include: improving relationships with healthy same-sex friends, community involvement, hobbies, spiritual and physical health, and working at improving their existing marriage.   While this method is certainly less instantly gratifying, it is almost surely significantly more rewarding in the future and the chances of attaining 'fool's gold' is substantially reduced.

"Compartmentalized relationships are like an out-of-focus movie"

- This analogy compares the projecting of positive and romanticized characteristics of brand new relationships to our brain's ability to correct an out-of-focus movie. Furthermore, it compares the inherent realistic flaws of this relationship to the flaws of the film when the projector self-corrects.

- This analogy helps explain to individuals why early phases of extra-marital relationships are often unrealistically blissful.

- Compartmentalized relationships

- Extra-marital affairs

- Projection

Our tendency to compartmentalize fantasized elements of our relationships with others is well chronicled in object relations and self-psychology literature. Nowhere else is this more apparent than with the client in the midst of an extra-marital affair.

Emotionally bereft for years, the patient finds himself emotionally vulnerable and susceptible to the slightest bit of attention that even partially fulfills his needs. In this compartmentalized form, the individual's ego attempts to rationalize their decision by creating a gestalt perspective of the other person. They generalize and project fantasized characteristics that their paramour may or may not possess. Equipped with this projected fantasy the patient, desperate to hold on to the part wish fulfillment, may be contemplating life-transforming decisions (i.e. separation or divorce) without rational regard for the consequences. Perhaps they have already made such a decision and are confused and despondent regarding the deteriorating nature of the once idealized relationship.

To aid in the interpretation of these dense theoretical concepts one might employ the following analogy: a valued supervisor suggested I compare a difficult relationship to a movie that is slightly out of focus. Our mind, having a tendency to form a gestalt, or whole-from-part-relationship, fills in the out of focus features we cannot make out. We fill in the screen projecting wished-for characteristics and creating an artificially sharper focus. As the initial passion fades from the liaison, so too does the idealized relationship. This is much like when the projectionist corrects the focus of the projection, and the imperfections are allowed to show through warts and all. We have a therapeutic duty to caution our clients of this phenomenon.

"Fanning the Embers"

- The painstaking task of fanning dying embers to ignite a fire is comparable to the difficulty of sustaining in-session learned marital therapy skills outside of session.

- This analogy can best be beneficial with couples, who, while motivated, have perhaps let their marriage deteriorate significantly before seeking professional assistance.

- Motivation

As one might expect, couples entering psychotherapy often wait until their relationship is in a considerably deteriorated state of affairs. As is human nature, they have allowed the dysfunctional dynamics of their relationship to continue for too long and now are consulting a professional as a means of last resort. When they finally enter therapy, they are feeling hopeless and burned out. They have in their own minds, while perhaps utilizing ineffective means, already tried countless times in a myriad of ways to improve their relationship. This often results in motivation that waxes and wanes between sessions. Grateful for the various methods the therapist gives them to utilize between now and their next session, they still find themselves putting forth a half-hearted effort.

I realistically assure them that their work will be analogous to fanning the embers of a dying fire. It will take hard and persistent effort, but if there is something there, it will eventually and exponentially ignite, resulting in sustained satisfaction. I sometimes briefly share a story of being a young boy watching a fire begin to die out in the fireplace. I have them visualize themselves on hands and knees blowing to near hyperventilation in slow steady breaths attempting to get the embers in the seemingly extinguished ashes to glow and once again ignite the remaining fuel. I tell them that it sometimes took a long time, but with continued effort I could most often reignite the fire.

"Analogy to objectify Chapman's Languages of Love."

- This analogy compares the 5 languages of love proposed by Chapman to varying size vessels or tanks in which to be filled.

- This conceptualization of Chapman's constructs provides a quick and visual representation to aid couples in focusing on those behaviors their partners require the most.

- Useful in helping marital partners specifically focus on what their counterpart requires to aid in maintaining self-esteem.

The popular Gary Chapman book, *The Five Love Languages,* has provided ample utility in aiding couples to focus on what types of behaviors are best used to illicit a positive reaction in their partner. In essence, the author posits five basic languages or set of behaviors that humans engage in to express love and caring. In addition, it is his hypothesis that, as a rule, individuals have a tendency to hierarchically arrange these languages as ones that they require or appreciate more than others. These specific languages are: acts of service, verbal encouragement, physical affection, quality time, and love gifting.

I have indeed found this elemental conceptualization quite efficacious in improving couples' communication, especially in aiding each partner to know, and then focus on, those behaviors most appreciated by the other while eliminating those behaviors their partners overlook.

Despite diligent efforts from their wives, most of my male clientele in marital counseling have yet to spontaneously pick up this literary gem (I'm guessing most are still struggling to get past the foreword of the *Men are from Mars, Women are from Venus* tome.) So I often summarize the author's findings by making the 5 languages analogous to vessels or tanks, stating that most of us arrange these languages or vessels according to size; the largest tank being those love behaviors we require the most of and the smallest vessel being the "language" we need the least.

I demonstrate using myself as an example. I say that my smallest vessel or tank is gift giving. Like most people, I like to receive one or two gifts a year, but little placards secretly placed in my briefcase, or messages made in shaving cream on the bathroom mirror are lost on me. I explain that this particular tank overflows quickly and the heartfelt sentiment is thus wasted.

On the other hand, my largest tank, which is seemingly never filled, is words of encouragement. In fact, it often seems that there is a small hole in it. In using this analogy it serves to make concrete those principles the author writes about, aiding the client in short order to focus on those specific behaviors that fill their partner's tank while refraining from standing over the small tank over-pouring non-essential love behaviors.

"I'll take my castor oil but do I have to say 'yum'?"

- Replacing a historically maladaptive behavior with a healthier substitute is akin to the ingesting of caster oil.

- This analogy is utilized in marital situations where a professional finally confronts one of the partner's behaviors as destructive and needs to warn the other partner against over-using the "I told you so" rejoinder.

- Increasing compliance in healthier behavior choices

Invariably through the course of marital therapy, one or both parties are forced to confront behaviors deemed unhealthy for the marriage. Usually said behavior will be the thorn in the side of the other partner who has spent countless hours to no avail trying to get their partner to see the destructiveness of this behavior. The therapist often substantiates the maladaptive nature of these discretions. Fueled by the professional's validation, the temptation of the oft-offended party to invoke the 'I told you so' rejoinder becomes almost palpable.

This of course does nothing for the offending party in regards to following through with the more adaptive behavior asked of them. Usually at this time I point out the self-righteous behavior of the other by exclaiming, "Sometimes one has to take their castor oil, but we shouldn't expect them to say with a smile 'Yum! This castor oil is tasty!'" This bit of injected humor usually diffuses the situation allowing both parties an out while also increasing compliance of the new behavior.

"Disengaging gears as a means to change habitual communication patterns."

- Reflexive and perpetuating tautological communication patterns seen in dysfunctional marital relations are analogous to gears engaging to make the apparatus go. Like gears, it takes two to complete the interaction, and if one simply does not engage, the other cannot move it forward.

- Used to help couples change habitually maladaptive communication patterns.

The nuts and bolts of what we try to accomplish in conjoint counseling often comes down to assisting couples in changing entrenched maladaptive communication patterns. These hostile and often tautological communication routines are so well ingrained that the couple easily reverts to this seemingly default position. Continuing to go around and around in some sort of ritualistic and fluidly dysfunctional dance, these patterns become almost reflexive. Most often couples identify two or three of these patterns, and if they cannot, they are soon revealed in session.

An example of one of these circular patterns may be as follows.

> **Wife** (responding to her husband's request for increased physical intimacy): "I'm just too tired in the evenings."

> **Husband:** "You wouldn't be so tired if you didn't talk on the phone to Barbara all day."

> **Wife:** "I wouldn't have to talk to her so much if you only listened and talked to me more."

> **Husband:** "I would talk to you more, but every time I offer suggestions, you never take them and then it's just the same problem all over again. So I just stopped."

> **Wife:** "Well I would not be so tired if you just helped around the house a little more."

Husband: "I could help more, but I have to work more hours, since you wanted to have Robbie (their second child)."

Wife: "I thought we both decided to have the baby."

As one can see, many different issues are being covered in this exchange without any active listening or attempts at resolution taking place. It is just one question fueling a hostile response that in turn elicits another accusation.

An analogy I use is to have the couple visualize when one of these patterns arises, is to imagine the words of the other as gears waiting to be engaged. The only way these patterns of maladaptive communication can be set in motion is if the words indeed engage the gears. I visually demonstrate this analogy by either drawing two rudimentary gears complete with teeth or arranging my fingers on both hands to demonstrate the interlocking principle. I show that the communication can continue only if both gears are engaged. If one partner stops or disengages, the motion stops.

For example, in the above exchange, the husband could simply have chosen not to respond to his wife's complaints of being tired with an accusation. Instead, he could have simply paraphrased her feelings of fatigue. Later on in the interchange, the husband could have avoided engaging in the new issue of the second child by stating a wish to deal with one issue at a time. Even if the other gear continues to turn, or in this case argue, the gear will only harmlessly spin in place unless the other gear engages.

After educating the couple regarding this principle, if the pattern arises over several sessions, I simply state that I think the maladaptive gears are engaged. However, where I think this analogy has real utility is outside of session. Couples have often reported in subsequent sessions how they could prevent a common argument from spinning out of control by simply stating this analogy.

"Creating a healthy marriage is akin to maintaining and restoring a classic automobile."

- This analogy compares the difficult task of nurturing a healthy marriage to maintaining and restoring a classic automobile.

- Marital therapy

One analogy I often use in marital therapy compares the process of sustaining and maintaining a healthy marriage to that of receiving a classic automobile (i.e. 1965 Ford Mustang) on the first day of marriage. The analogy posits that while this car is not in pristine condition, it is in good working order, contains all original parts, and has the potential to be restored to mint condition. If accomplished this would be a sound investment, the car being worth much more than when initially received. Does one simply do the required maintenance that keeps it in its originally received condition? Or does one store it in the back yard vowing to make the time to restore it to mint condition while distracted by day to day living (career, kids, alcohol, golf) thus allowing the grass to grow high around the ever-rusting classic?

On the other hand, does the couple make the necessary commitment to continually restore the car to showroom condition? I have found this to be an efficient way to assess the current state of my client's marriage while at the same time being able to communicate the corresponding amount of work the couple will have to undertake to progress to the desired state of marital functioning.

It is important to demonstrate here that not all analogies work for all people. While painstakingly relaying this analogy for editorial approval, my girlfriend emphatically stated, "But I would want a new car!"

"Marriage is often like a bank account."

- A balanced bank account, complete with a ledger of deposits and withdrawals, is akin to the sustaining of a healthy marital relationship.

- This analogy is useful in explaining to clients the need for proactive, respectful marriage building behaviors (deposits) and the reduction in negative interactions and transgressions (debits or withdrawals).

- Forgiving of old transgressions (debts)

- Increasing compliance for newer healthier behavior

In a straightforward and probably not too original analogy, it is easy to compare interactions between couples to a bank account. Positive experiences and communications are analogous to credits or deposits to an account, and conversely negative interactions and transgressions can be seen as debits or withdrawals. Conjoint research has shown the significance of day-to-day positive interactions as imperative to a healthy marital relationship.

One unique use came in a treatment episode where a husband, significantly impaired by years of depression and substance abuse, was emotionally absent, leaving his wife to fend for herself in all aspects of the relationship. Now sober, with his depression adequately treated at a sustained sub-clinical level, he was for the first time able to truly communicate with his wife. Woefully inadequate in relating to his wife after years of depressive malaise he was now able to perform the rudimentary functions of communication. He now looked at his wife while she spoke in an active listening posture and was able to accurately paraphrase the content of her communication. Slowly he began to approach equality in the division of household chores and responsibilities. Still, there was much hostility between the two; he feeling not appreciated for his newfound efforts, she still bitter at the years of skewed responsibility.

As a therapist I could validate both parties' feelings. Still, I was afraid he would become frustrated, give up and return to his emotionally bereft ways, only increasing her already righteous hostility.

I utilized the analogy of the bank account during one of their repeating maladaptive patterns of communication described above. Furthermore, I suggested that years of withdrawals without corresponding deposits had left his account significantly overdrawn and in arrears. I validated him for his current status of making consistent albeit minimal deposits. We mutually attached a figure for how far he was in arrears and the amount of the deposits he was now making. This momentarily broke through the existing hostility, even injecting a little humor into the situation. I offered to her the possibility of being a more patient banker, trading her hard line stance to one of increased patience. Much like the benevolent collection agency, she might look past the years of delinquency and more to the small consistent deposits. I even asked her to evaluate the possibility of "writing off" some of the debt. I could encourage both. As he became more versed in the ways of making deposits they were likely to increase in amount and frequency. As she was able to recognize the effort he was making, she would be able to look more favorably at his account. This proved beneficial even outside of session as they often reported that they joked about the amount of emotional "deposits."

"Nurturing a relationship is akin to nurturing a garden."

- The sustaining of a healthy marriage is analogous to the planting, maintenance, and special attention one needs to provide in order to achieve a healthy garden.

- This analogy is useful in couples understanding that a marriage needs a good foundation (respect and genuine acceptance of each other, value and expectations clarification), maintenance (structured date nights) and sometimes-special attention.

Similar to the bank account analogy, the maintenance of a healthy relationship can be seen as akin to nurturing a garden. Not only is regular maintenance, such as watering a necessity, but it also periodically requires extra nutrients. This is similar to healthy day-to-day interactions being like watering, and quarterly special date nights or weekend getaways being viewed as special nutrient supplements. A patient once put a humorous yet truthful spin on this analogy: recognizing that she was a bit on the high maintenance side, she identified herself as an exotic flower, thus needing special food and a little extra watering.

"Marital change is often like a dartboard."

- This analogy compares the practicing of new healthier behavior to aiming at a dartboard. Sometimes the other partner is initially expecting that the partner hit a narrow target.

- This analogy can be useful when one partner is having trouble performing a new behavior.

- Reinforcing the approximation of a new behavior as a means to shaping and increasing said behavior.

Rarely do couples enter therapy at the first sign of communication difficulties. Usually there have been years of entrenched and maladaptive patterns. Individuals have spent sometimes decades attempting to get their spouse or partner to understand basic needs or desires and have developed more than a modicum of hostility because their wishes have gone unheeded. Finally, at the perilous risk of their marriage terminating, the couple agrees to seek professional help. Through the process of counseling, the predominantly offending party begrudgingly humbles themselves and recognizes the errors of their ways. Often this is the result of gentle therapist confrontation.

Once the recognition has been acknowledged, the arduous process of changing these dynamics and behaviors can begin. Sometimes in these situations, the partner, finally validated after all these years, gets a little overzealous in their expectation. They are going to make their spouse pay for their previous transgressions. This payment comes in the form of berating or chastising them after each slip-up. Fueled by years of hurt, frustration, and disrespect, they even do this when their partner is attempting the desired behavior. Granted the partner's attempts may seem a tad inefficient in this neophyte form, but to the clinician the genuine attempt is apparent. Confronting this dynamic maximizes the reinforcement of the new behavior and significantly reduces the possibility of the partner giving up.

Using the analogy of a dartboard, I delineate levels of the desired behavior from a sure miss (the black outside of the dartboard) to a reasonable approximation (the larger slices of the board) to a bull's eye (the exact and for years wished for response or behavior). I explain to the partner that to a certain extent they are asking their partner to throw a bull's eye after just learning the basic rule of darts. I explain that they are giving them a bit too narrow of a target to aim at and there must be some reason they are doing that. This often leads to the frustration and hurt they have felt for so long and the uncovering of resentment for their partner not trying harder long ago. Although we discuss the need to let go of these paralyzing feelings, it is also important for the other partner to really hear and understand the years of hurt the other has legitimately felt. I often find that the offender, if hearing this absent the persistent haranguing, will increase his or her motivation to work harder to hit the narrowing target.

Analogies for Parenting

"The cops have the power of the law behind them."

- This analogy compares the police officer being able to enforce the law more efficiently because of underlying laws, to a parent's ability to better parent if they have a well thought out behavioral contingency plan with natural consequences.

- Used in parenting situations where parents make instantaneous consequences motivated by frustration and anger without the aid of well thought out consequences. Also useful in situations where a parent gets in frequent debates with a particularly manipulative and precocious child.

- Natural consequences

A majority of children seen in psychotherapy today have at the core an oppositional behavioral problem. Most involve neurological predispositions and exacerbating environmental conditions necessitating a multimodal approach. Still a significant amount of cases involve nothing more than faulty parenting approaches, specifically situations where the child-parent interaction is fraught with hostile emotional interchanges that leave both parties exhausted and angry.

Often parents bring their children to therapy believing there must be some serious neurological disorder or 'deep-seated-anger' problem for their children to behave in such an atrocious way. These parents describe not being able to get their child to mind or listen. They will assure you they have tried everything from grounding to spanking but nothing seems to work. They will describe situations where the child will directly disobey them causing them to capitulate to get them to perform the desired behavior. These bribes are only a temporary solution to exacting more lasting behavioral change.

Flustered by the child's surprising unwillingness to comply, the bewildered parent, caught off guard, relies on overemotional responses, often exacting severe punishments. Following through with these consequences is another story, thus rendering them useless. During the initial clinical interview, the parents express further astonishment that this behavior only occurs at home and not in other environments such as in school or at the homes of friends.

Inconsistent and faulty parenting practices often create this behavioral problem. The parent may feel they will 'damage' their children if they do not let them 'express' themselves. Other parents may have a restricted repertoire of creative approaches, or they are simply repeating the dysfunctional patterns of how they were raised. Most approaches to increasing behavioral compliance involve an authoritative parenting style (one that places a premium on immediate compliance while at the same time providing a socio-educational explanation for the request) along with developmentally appropriate and consistent consequences that reflect some forethought. To emphasize the importance of this notion and to provide a real life application of the procedure I relay the analogy of why a police officer does not have to get mad at us during a traffic stop.

In this analogy, I often ask the parent why a police officer does not have to become upset at us during most routine traffic stops. I then go on explaining that this is true even if we choose to get unreasonably worked up ourselves. After they struggle for a while with this scenario, I offer the explanation that the reason the officer does not have to lose his temper is that he has the law behind his actions. He does not have to make arbitrary decisions fueled by his emotional state at the moment. He simply has to enforce the well thought out integrity and history of the various laws. I actively demonstrate a scenario of a routine traffic stop for speeding. As the officer, I am apologetic; "I'm sorry Mr. Seaton you are were going 55 in a 40mph zone which means you owe (glancing down at his fee book) 150 dollars." As the driver,

I progressively become more irrational and upset culminating in me tearing up my ticket. Again, I demonstrate the apologetic law enforcement personnel: "Mr. Seaton, I see you have torn up your ticket; well we have a copy in the computer which contains your court date. I am afraid if you do not show up you may risk greater sanctions. I'll tell you what, I have to go now but there is a phone number in the blue pages of your phone book where you can get a copy of your citation."

Obviously, this exaggerated illustration is to emphasize the in-control approach a parent can have in a difficult child-rearing situation if they have well thought out in advance developmentally appropriate consequences. To further illustrate the importance of having this structure in place as essential to deescalating emotional power struggles, I propose the following: "Can you imagine if the officer didn't have the law behind him?" He would be left up to his own devices to dispense judgment at the very moment he was overly influenced by his current emotional state. Perhaps he had a late night, a disagreement at his previous stop, or just a long day. I continue on this path playing the driver as if I were the verbally precocious youngster the parents have brought to my office: "Hey, you stopped my friend last night going the same speed and you only fined him 100 dollars." I ask the parents to imagine where many of these interactions would end up if the officer had nothing except his brute force, weaponry, and intimidation to resolve disputes.

I choose a male police officer as I have found this analogy has good utility with fathers. Often in the therapeutic situation the father favors a harsher approach, which although temporarily adequate, does not scientifically lead to more internally based compliance. This example models for these fathers a compassionate yet firm male authority figure, while illustrating the need for immediate compliance to often passive mothers.

"Give 'em one bag of sugar at the border and they won't search your car."

- Giving one or two harmless bags of contraband while keeping others is comparable to convincing adolescents that giving their parents some personal information may actually decrease their parent's nosiness.

- This analogy is often used with overly reticent adolescents and probing parents in an attempt to open lines of communication.

- Rapprochement-autonomy in adolescence

- Value clarification

- Formal operations

Perhaps there is no more challenging patient than the reticent adolescent. Despite attempts at open-ended questions, the recalcitrant teenager meets even the best of these with the unenthusiastic, "I don't know" or "not really". If we as clinicians, expertly trained in eliciting information get frustrated within an hour, one can only imagine how difficult it must be for parents who go through this communication block on a daily basis.

Much of what we do in adolescent psychotherapy is aid the client and their parents in striking the delicate balance between maintaining the teenagers burgeoning individual identity while at the same time protecting the integrity of the needed boundaries still inherent in the parent-child relationship.

Adolescents, bombarded by hormonal and cognitive changes, find themselves embarrassed by the content of their thoughts and experiences. This narcissistic phase of beginning formal operational thinking lead them to irrationally believe that these feelings are uniquely their own and could never be understood by their parents. Thus they invest inordinate energy in offering as little information as possible in an attempt to keep their parents at bay. Suspicious and rejected parents, who have seen their relationship with their child deteriorate from holding hands sharing ice cream at the mall to now only communicating through closed bedroom doors in seemingly a matter of months, become Gestapo interrogators in an attempt to regain this previous bond. The less the teenager offers, the more the parents turn up the heat and frequency of the questions. The more they ask, the more the adolescent recoils. Thus is the classical impasse of rapprochement in adolescence. When it is evident that this is occurring in session I often share this analogy: Growing up relatively close to the Mexican border, our family made frequent trips for leisure. During these excursions, there was the inevitable bartering for goods found much cheaper south of the border. For reasons that now escape me, our family's habit was to purchase multiple bags of sugar and shrimp. Our father would load the trunk with the bounty except for a couple bags of each item that he would then place in the front seat. As we stopped at the border crossing station, the officer would ask us if we had any produce or plants. My father would reply, "Well we have these bags of shrimp and sugar." The guard would dutifully confiscate the items and would send us on our merry way. We then passed into the United States carrying a majority of the loot.

After sharing this analogy with the adolescent, I then ask the teen what would have happened if upon reaching the border my father would have responded to the guard's request by uttering: "I don't know. Nothing I guess" or what would the guard have done if my father had answered in a hostile "I don't know. Leave me alone!" What if he would have responded with nervous trepidation, obviously trying to conceal information? I usually role-play these possible responses with exaggerated effect. The astute adolescent usually quickly surmises that the border guard would have probably brought out the dogs and stripped our car bare. I use this analogy to illustrate that usually parents want just the basic knowledge of what is going on in their lives. For the most part, they want to know if the child is making good choices as it pertains to friends, education, drugs, alcohol, and sex. I hope to encourage and convince the adolescent that by giving their parents a little pertinent information, they may not have to endure as much probing. This analogy is usually best relayed in front of the parents, so they can understand their part in the dynamic as well as assure their teen that they realize they are going to leave out details.

"Remember that "I Love Lucy" episode, the one with the cakes on the conveyor belt?"

- This analogy compares the classic sitcom episode where Lucy is decorating cakes to the often-fixed timetable of parenting.

- This analogy is often used with parents feeling pressed for time to exert their influence on their late adolescent or young adult children.

- Identity formation

- Early representations of parenting sustaining the child throughout development

Most of us recall the now classic episode where Lucy is decorating cakes that come through a conveyor belt. At first, things are running smoothly and Lucy performs the required task of decorating the cakes, placing them in a box and taking them off the conveyor belt. It is only when she makes a simple mistake and wants to correct it before boxing it up, that typical Lucy hi-jinx ensue. She realizes that there is little time to correct the mistakes before the next cake comes along. She tries to push the approaching cakes back up the conveyor belt, but soon they simply pile up along the belt. Eventually she resorts to sloppily decorating the cakes and throwing them in boxes so she can just catch up.

For some reason this scene spontaneously came to me in session while talking to parents of a 19 year old who was going through some moderate adjustment problems. These parents, perhaps feeling some guilt regarding their busy career-achieving lifestyle during their child's formative years, were trying desperately to control various aspects of their now young adult's life. Needless to say, this was running up against a great deal of resistance on their child's part. At this time, I relayed that the cake and the conveyor belt seemed to be analogous to raising children. The cake, like the child, seemed to be a rather plain pastry that we could decorate or exert our influence upon any way we wanted. Time and developmental stages, like the conveyor belt, seemed to move at a leisurely if steady pace, giving the appearance that we had ample time to complete the process before neatly boxing up our perfect child to send off to be a productive, happy member of society. It is only when the child nears the end of adolescence that we realize we will not be able to influence or decorate our child the way we had hoped. We try everything to stall the process or hastily cram our influences into their already somewhat stable identity. Nevertheless, like the Lucy Show cakes on the conveyor belt, this often leads to disastrous results.

This analogy seemed to break their hopeless focus, allowing them a broader perspective. They were then much more open to various object relations' theory interpretations explaining a child's early internalizations of various parental influences as stable guiding forces throughout the child's life. Trusting the therapist's interpretations they came to the conclusion that while not being able to 'box' up the perfect child, they realized they had probably been as Winnicott would say, 'good enough' parents.

"Raising adolescents is akin to placing guardrails on a highway."

- The delicate process of raising an autonomous yet responsible adolescent is analogous to the proper placement of guardrails along a long and curved highway.

- This analogy is best used in aiding parents and children to set proper boundaries, which insure safety, yet foster autonomy and healthy identity formation.

- Boundaries

- Autonomy and identity formation

- Parental letting go

An esteemed supervisor offered this analogy when discussing a case of mine involving helping parents set proper boundaries with their adolescent. He had me visualize a curving highway with some perilous drop-offs. To aid the novice driver in negotiating this particular stretch of road we would need to strategically place the guard rails in such a way as to provide the driver with a sense of mastery while insuring his ultimate safety. To place the guard rails too close together would falsely give the driver the feeling that he was adequately maneuvering the course when in reality he was simply being guided by the placement of the rails. For all intents and purposes, he could simply take his hands off the wheel and he would still be completing the course. To not have any guard rails at all would surely result in the driver swerving out of control, and perhaps over the embankment itself. The optimum placement would allow the driver enough control to adequately grasp the feel of each individual turn. However we might place the rails at those most potentially dangerous curves insuring that if our driver swerved too far he would meet just enough resistance to nudge him back toward the straight line of the course.

This particular analogy allowed me to confront the parent's restrictive boundaries in such a way as to not significantly damage the tentative rapport we had established. It also helped the adolescent see the need for certain realistic boundaries. This analogy demonstrates the importance of autonomy in the adolescent's identity formation. They must feel their choices are internally driven and not simply a function of their parent's wishes. The parents in turn must see that failure to let their teen steer through the trials and tribulations of adolescence, trying on different identities, will preclude them later on from making authentic and autonomous decisions.

"More than one adult eagle in the nest"

- Delayed emancipation of an adult child is analogous to the presence of more than one adult eagle in a tiny nest.

- This analogy is best used in situations where there is an under-producing young adult child who remains in the home draining precious resources and becoming increasingly dependent.

- Enabling

- Codependence

- Emancipation

Another helpful analogy one might use regarding issues of adult child emancipation compares the adult child remaining in the house to a newly adult eagle still living in the nest. Good comparisons can be made using the gathering and division of food to the client dealing with this prickly situation. The mother eagle working hard to bring food home to the now able but untested adult eagle can be compared to the parent and the under producing young adult. The mother is likely to feel ambivalence or even anger at still needing to regurgitate the food for the eagle that barely lifts a wing. Analogies can be offered to demonstrate the cramped feeling of all adult eagles in the tiny nest, and the predisposition for aggression as tensions mount in the too restricted space.

The therapist, in order to get at the usual root of enabling (guilt for fear) can describe the nest as high upon a cliff, and the mother as frightened for the eagle to fly from the nest. What if they are not able to achieve flight? You may be able to assist the patient to see the instinctual choice the grown eagle needs to make as it forcefully nudges the younger eagle out of the nest. To not do so results in an enabling paralyzation of the youth who will only become increasingly dependent. Visualizing grown eagles fighting for room in the tiny nest usually brings home the urgency of the situation.

Despite the inherent urgency, growing tension and cramped quarters, parents finding themselves in this situation feel hopeless to exact change or elicit increased motivation in their young adult. Usually at the root of these situations is fear and insecurity. The parents fear their child is ill equipped to enter the world as a productive citizen, and the child, perhaps growing up in an environment of passive enabling, for the first time truly doubts his chances. To assuage the parents fear and to send an implied message to the young adult that they can indeed make it, a hierarchal plan is devised to transition emancipation and to foster successful flight. An example may be setting up a target date 6 months away. During this time, the adult is asked to contribute a progressively increasing amount of rent until the target date. The parents may provide a matching incentive to any savings the child may accumulate. The importance of enforcement of the target date is paramount in these cases.

"The runt of the litter hangs around."

- Continuing to enable a dependent adult child is similar to a lion pride allowing adolescent males to linger too long.

- This analogy is useful with families dealing with emancipation issues involving a dependent adult.

- Enabling

- Emancipation

- Parenting

- Separation-individuation

- Rapprochement

One of the more challenging situations a clinician faces is helping parents devise an emancipation plan for an adult child they have enabled for many years. Persistently complaining about their child's laziness and unappreciative attitude, they are nonetheless resistant to consistently enforce realistically progressive boundaries geared to promote emancipation and independent living. Usually at the core of the parent's failure to demand their child face the world is fear and guilt. Perhaps preoccupied with their own self-absorption in maladaptive circumstances (dysfunctional marriage, career, substance abuse), they feel they did not equip their child with the proper set of coping skills needed to succeed. They fear that if made to go out on their own, the child would not be able to make it and possibly face physical harm.

It is imperative to convince the parent that continued enabling actually sends a strong implied message to the child that the parents do not feel they are capable, only reinforcing the young adult's low sense of competency. It is often quite difficult to get the parents to recognize their own complicity in the situation. They are naturally defensive, assuring you that they have provided their child many chances. To confront this resistance, it is quite useful in these situations to take a family systems approach in explaining the need for separation.

Like many of the analogies in this book, this one spontaneously came to mind while involved in a therapeutic exchange with a parent struggling with this dilemma.

A young cub, one of a litter of three, was born to a lion pride on the African Sahara. This cub was much smaller than his siblings. Scrawny by comparison, he remained the runt of the litter throughout his young life. Although the runt, he had an uncanny ability to survive in these harsh circumstances, doing just enough to gain the minimal sustenance required to remain with the pack. His siblings continue to gain in strength and size, constantly making him the mock victim in their instinctual play, knocking his legs from underneath him and simulating the strangling chokehold they would later use during the hunt. The cub persevered, but it was quickly becoming apparent that this lion was developmentally impaired.

Then one day the cub, innocently exploring his terrain, finds himself separated from the pack. Despite the continuous calls from his mother, he remained lost, and as is the harsh survivalist way of the pride, they had no choice but to move on. The scientists documenting the pride stated that the cub had little chance of surviving in this harsh environment, and would most likely be killed by rival predators.

Somehow, after almost two weeks, the emaciated cub miraculously reunited with the pride. The seemingly joyful reunion is tempered by the now exponential gap in size between himself and his siblings. They were becoming adolescents, beginning to display the small tufts of back hair that would later become the full-fledged mane of the adult male lion. The pride had also expanded during this time, adding hungry new cubs dependent on the pride for their existence. Soon after, the siblings are ousted, earmarked for their own pride, or perhaps a later triumphant return to supplant those who had ousted them. In contrast the pride seemed almost confused by the anomaly of the lost cub's survival. Against their natural path of instinct, the cub now enjoyed a unique place. He was free to fraternize among the male leaders. Not sure what to make of this larger cub, they treated him as such, instead of a potential rival. This expanded adolescence lasted during the rainy season and into the dry months.

This languorous existence enjoyed by the cub is shattered in an instant while attempting to take his usual feeding position at a recent kill. As he begins to feed along side the dominant male, the same male that he was playfully jostling with only days before, the larger lion turns his head in menacing fashion, bares his large canines and bellows a ferocious roar. The startled cub leaps back three feet before slowly returning to the quickly disappearing carcass. This time the older lion attacks the cub, exacting vicious claw marks, leaving no doubt of his intent. The cub was no longer welcome at the kill, no longer allowed to partake in a hunt in which he was not a participant. The confused and injured cub saunters over to the females in the pride for support and comfort. The females, following the lead of the dominant male, repeatedly rebuff the now adolescent cub's attempts for solace. Within a 5-minute time span the young lion that had survived so much was transformed from an overgrown cub, the pride mascot, to a potential threat to the prides very existence. He was banished from the pride forever.

As the viewer watched the still thin and undersized lion wander off into the darkness, the scientists documenting these events stated a grim prognosis for this supposed king of the jungle. As a rogue lion he was perhaps destined to a life on the periphery subsisting on carrion or small prey if he was to survive at all. The nature show ended with a cryptic statement from the narrator stating that in fact they never knew for sure if they ever saw the lion again.

When viewing the program, especially as the lion disappeared into the dark horizon, a multitude of ambiguous feelings struck me. A feeling of over protectiveness came over me. He was not ready to hunt for himself. Anger also emerged toward the pride. Didn't they realize what he had been through? Why hadn't the pride trained him better, making him part of the hunt more often? I could not relate with the harsh treatment they gave him at the kill. Didn't they feel guilt as they expelled him to who knows what fate?

The narrator chided my anthropomorphist thoughts and feelings, explaining that had the cub been allowed to stay, the delicate dynamics of the pride would be significantly altered, and would become detrimental to its very existence. He stated that during the season of drought there was simply not enough food to ensure the survival of the pride, including the new cubs. In addition, when the inevitable coup took place with a new set of dominant males ousting the old guard, the cub would be banished anyway, perhaps killed by the reigning victors. Even if he did live, at this point his skills would be even less honed. A Darwinian pall washed over me as I struggled with my human feelings and obligations. Lacking the workings of the frontal cortex, the animal's decisions become instinctually correct and simple in their execution. Our choices regarding our offspring are an inherently more difficult dilemma.

"Dilator"

- This analogy bluntly compares the possibility that sometimes we can choose the size of the dilator being inserted but not whether it will be inserted to the acceptance of life events we cannot control.

- This analogy is often used to aid a patient in diverting often negative and further destructive energy in situations such as marital dissolution, business failure or personal feud.

Many individuals find comfort in the norm of their status quo situation and any necessary change to that condition often brings on a tenacious fight to restore equilibrium. There are unfortunately situations in life that call for some resignation in order to avoid the exhaustion of fighting an impossible battle. We as clinicians often need to guide them toward surrendering with some sense of dignity. The dissolution of marriage, complete with loss of time with children and requirements of paying child support, causes anger, resignation and hostility. The devastating impact of this situation is no less drastic than any unforeseen business failure or familial abandonment. Common among these situations is a demoralizing feeling of unjust betrayal that is difficult to accept. These individuals engage in pointless legal battles often without merit. Our goal as therapists is to help them negotiate this unavoidable need for resignation and modification without too much of a threat to their already damaged sense of self-worth. Frequently these individuals' actions are only making their situations worse, bringing additional punishment that they don't deserve.

Some individuals, men in particular, need a somewhat blunt analogy as a means to confront difficult choices ahead. I have at times truly empathized with their feelings but feel the need to tell them frankly that although they may not have the choice to decide whether a dilator gets inserted rectally, they do (by choice of their behavior) have the choice of what size dilator gets inserted. There are many situations where this analogy fits (so to speak, no pun intended. I thought about using the synonym, applied, but it did not lessen the double entendre). Patients upon termination of therapy repeatedly cite this analogy as a turning point in getting them to refocus efforts to minimize damage and consider the possibility that the urgency of the current situation is temporally transient, and likely to ease significantly over time.

"The best parenting approach is analogous to a two-headed benevolent dictator."

- Authoritative parenting is analogous to a two-headed benevolent dictator.

- This analogy is utilized to introduce the most effective parenting style. It is often used when two parents vary widely in their child-rearing and discipline strategies.

- Parenting and child-rearing (autocratic, authoritarian, authoritative, permissive)

- Splitting

- Conscious building, frustration tolerance, and delay of gratification

Frequently we observe that parents in the same family vary widely across parental approaches, perspectives, and continuums. We have the authoritarian parent teamed with the permissive. The passive parent demonstrates her laissez-faire approach to compensate for her harsh counterpart. He in turn overcompensates to counter her move becoming even more unyielding and restrictive. In full observation of the child, continued undermining of parental decisions causes an ever-growing chasm of inconsistency. Needless to say, even the relatively inexperienced child quickly learns to split between parents, increasing maladaptive and manipulative skills that may generalize to other environments as well.

Then there is the overly democratic parent who places the child in equal status with the adult. With these parents any attempt at limit setting, a fundamental prerequisite for conscience building, ability to delay gratification, and social cooperation, is erroneously perceived as not treating the child with respect. Furthermore, it is falsely believed that it may later hinder the child's creativity. This child-rearing behavior has been carried to such an extreme that I have personally witnessed a graduate professor and his physician wife unable to get their 6 year-old to eat without nightly playing the chase scene from *Pee-Wee's Big Adventure*. These parents come to my office incredulously surprised that life with their children is a series of bribes to achieve acceptable behavior. This is not all that difficult when the bribe is a 25-cent gumball token. It becomes a little more challenging when the adolescent insists he receive the DVD system in his new SUV.

When I run into these parents I state that the most effective type of parenting is akin to a two-headed benevolent dictator. In essence, this is a concept that best represents the authoritative approach to child rearing. It is autocratic, and one where the children are definitely of lower power status. However, in this type of dictatorship, the leader is benevolent, disseminating educational reasons on why one has to follow these tenets. The two-headed aspect explains the need for the parents to put up a united front, using phrases such as "we have decided that you must earn the privilege first." At its optimum use, the child should not even know which parent came up with the various consequences.

"Erasing the lines a little at a time"

- This analogy compares weak boundaries in parenting to erasing the boundary lines of any phenomena. While barely perceptible and innocuous at first, over time, drastic changes in acceptable behavior occurs.

- This analogy is helpful for parents struggling with boundary setting

- Boundary setting

Frequently a parent who demonstrates a problem with firm boundary setting finds themselves raising a verbally precocious child gifted in the ways of debate and manipulation. By means of pleading, arguing, or negotiating these children simply overwhelm the beleaguered parent, causing them to dilute their parental demand.

It is helpful to demonstrate to these parents the effect of erasing boundaries just a little at a time and how the cumulative result can be significant. The therapist can demonstrate this by first marking a boundary on the floor of the office. The therapist can then show by erasing and remarking the boundary how each argument serves to stretch the limit. After several instances, the significant altering of the boundary is quite evident.

For example, this is similar to a child who extends his 8:30 p.m. bedtime by begging to stay up until the next commercial break of a television program. Soon after retiring to his room, the child cries out that he needs a drink, or states he needs to use the bathroom. The parents acquiesce, and soon this becomes a nightly occurrence. Now however, the child additionally requests a bedtime story. Before long the original 8:30 bedtime is now consistently 9:15 p.m.

I first came across this analogy as someone was explaining how civilization changes. Big revolutions seldom significantly change cultural mores and values. However, he explained that smaller, often undetected pushing of societal boundaries insidiously alters what is accepted.

For example, the African-American influence on rock and roll music of the fifties was not readily accepted by the dominant culture of that era. Instead, white acts such as Pat Boone crooning tame versions of black originals lead to the progressively provocative renditions of Elvis Presley and Jerry Lee Lewis. Finally America could now accept Little Richard singing his original tune initially made popular by Pat Boone.

"Entrenched power battles with a child are like brakes that lose their padding."

- The consequences of entrenched and protracted power battles with difficult children are comparable to the loss of brake padding resulting in increased friction and irreparable damage.

- This analogy aids parents hopelessly stuck in singular and restrictive parenting patterns with difficult children.

- ADHD

- Children with difficult temperaments

- Childrearing styles

Parent-child dynamics, especially those involving special needs children, can become so protracted that they reach pathologically harmful levels to both parent and child. A good example of such an interaction is a parent raising a child with Attention Deficit Hyperactivity Disorder (ADHD). The inherent inability to contain impulses, difficulty in learning from previous consequences, and propensity towards destruction makes parental supervision of these children an exhausting endeavor.

The parent is constantly finding themselves telling the child "no", and become overwhelmingly frustrated that their child cannot seem to comprehend their demands. For the most part the child with this exacerbating condition does not act out with malicious intent; it is simply a product of the minor brain dysfunction characterized by ADHD. The parent often misinterprets these behaviors as volitional and disrespectful becoming increasingly punitive in an attempt to get their children to behave. The child, wanting to please their parent but finding this difficult, eventually feels helpless in their attempts. They may become frustrated and increasingly argumentative. The parent, bound and determined not to give in, becomes more entrenched in a singularly and ineffective mode of behavior management. The child resists with equal force and soon one sees in the therapeutic office the unpleasant result of such a dynamic.

It is important to join with the parent in their need to make the child behave. However, it is paramount to get parents to understand that they are pathologically stuck in an ineffective mode. The goal is to help the parent expand their repertoire of behavioral management techniques, while educating them about real organic variables inherent in the equation. An analogy I offer compares the interaction between child and parent as akin to how a brake works.

I offer a rudimentary explanation of how a brake functions, comparing the stopping function to stopping a child's inappropriate behavior. At this point, I explain that between the two metal brake plates there is a protective piece of padding which keeps the plates from experiencing an inordinate amount of friction. The padding lasts for a specific amount of stops, before becoming worn and manifesting a squeaky noise. If worn too thin the amount of friction and heat increases to often-unbearable levels. If no padding exists between the two brake plates, the stopping function becomes essentially metal against metal. This process can leave irreparable and permanent grooves in the plate and on the disc.

Most parents are adroit at recognizing how this analogy describes the interactions with their own children. They are able to quickly gauge the deterioration of padding (this padding being akin to natural and instinctual forms of love a parent and child have for each other) between the two parties. The friction and heat in this case is the increasingly hostile interactions between the child and parent. The grooves, the ultimate manifestation of damage, can be seen as potentially irreparable emotional harm. Understanding this analogy often allows parents to be open to expanded methods of behavior control and the potential organic and not necessarily volitional causes for the child's behavior.

"Parenting is akin to running your engine at higher rpm."

- In this analogy, the increasing and sometimes overwhelming demands of parenting are compared to running an engine at higher rpm. Thus, there is a need to replenish the relationship between adults even more.

- This analogy can aid new parents who find themselves exhausted and neglecting other aspects of self and the relationship.

- Date night

As many of us know, the advent of parenting brings about many new stressors and responsibilities to our lives. It appears that the day is shorter and all of our nervous systems must run faster and more efficiently to prevent complete breakdown. In a marriage this often goes unheeded and the first things to go are those activities that replenish the relationship. This does seem to run contrary to intuition. If something runs much faster, it should require increased maintenance and lubrication. That is why I often compare this new stress to forcing the engine to run at more rotations per minute. Now more than ever there is a need to lubricate the engine and keep to the maintenance requirements.

Analogies for Behavior Change

The next series of analogies has to do with how we can help our patients/clients conceptualize the therapeutic process of behavior change. Often a client, after painstakingly detailing his or her presenting problem and concomitant social history, will turn to me during this initial session and ask, "What I do now Doc?" To them, their situation seems so overwhelming and convoluted they have no conception of how they will reach any degree of resolution. It is difficult for them to believe that anyone can help them untangle their psychological mess. Initially, they simply cannot fathom our expertise and training in these matters. On some level they are ambivalent, sensing that we must have some modicum of knowledge but they require at that desperate moment at least some tangible evidence that we understand their problem and better yet have a tactical approach on how to solve it.

This is an initial critical juncture in the development of therapeutic rapport. If we give them some overly simplistic answer, it will strike an inherently discordant note and they will doubt our commitment or expertise. If we make the process sound too arduous or burdened with theoretical jargon, they are likely to have difficulty engaging and perhaps not even return for the second session. Often a well thought out analogy providing a beginning, middle and end can serve this purpose.

"The radiator and the cooling of the engine"

- The radiator, a mechanism that regulates and cools the engine, is akin to the development of ego competencies as a regulating mechanism of the self.

- Used in cases where patients are suffering from character deficits or extreme affective dysregulation.

- Affective regulation

- DBT

- Object relations

For a myriad of reasons many of our patients, especially those that are manifesting characterological deficits, do not have the required psychological processes to deal with the life stressors presented them. Whether we term this ego capacity, faulty cognitions, self-esteem regulation or coping mechanisms, these individuals are constantly finding themselves in a state of psychological overload unable to deal with the stress they are encountering. Their problems are akin to a radiator that cannot meet the cooling demands of the engine. This particular analogy comes from a brilliant patient of mine who suffers from Asperger's Syndrome. While this individual readily admits a difficulty with spontaneous social discourse, she proves quite adroit at the written language or as she termed it her "ability to turn a phrase". Her analogy came in the form of an email she sent attempting to explain her feelings, something she could not express to me in her session.

She provided initially a rudimentary explanation of how a radiator works to cool an engine. First the radiator is filled with a cooling agent that is dispersed throughout the engine by a series of hoses. The movement of the cooling agent cools the motor as the heat demands increase. A thermostat is included to gauge the temperature and to decide how hard the radiator has to work. She felt that her radiator was just too small. She surmised that there were possibly 4 approaches to deal with this radiator deficiency: one could ideally increase the capacity of the radiator, one could lower the amount of heat the engine creates, one could introduce other mechanisms to cool the engine, or one could lessen the effects of overheating the engine. Reading this analogy before our next session provided a flood of analogical ideas, the central concept being the radiator as akin to a central mechanism i.e. the ego, as a means to regulate a multitude of psychological functions. Her analogy is another great example of how therapeutic analogies can often work as an implied and mutually understood treatment plan between patient and therapist and in fact was used in this way throughout her treatment.

To follow this analogy in general terms, let us view the radiator as equivalent to a central psychological regulating mechanism. Much like the ego, this mechanism plays a significant factor in self-esteem regulation, decision-making capabilities, stress management etc. In this analogy, the heat of the engine can be seen as any stressor that approaches the self. By becoming more adept at recognizing those stressors that tax the system, the patient was able to regulate how hard she pushed herself in certain situations. Learning techniques such as stress management strategies, or asking for assistance or support was her way of increasing the capacity of her radiator.

"Behavior change is like learning to play chess."

- The process of behavior change is similar to the process of learning and playing chess.

- This analogy aids the client in observing that cognitive insight often precedes emotional insight.

- Unconscious motivation

- Cognitive reframing

- Discriminate stimulus

Intellectual or rational knowledge of a desired behavior change usually precedes our emotional understanding of how to consistently execute said behavior. During the therapy process, we aid our clients in understanding that their maladaptive choices are rooted in ingrained emotional dynamics just outside their conscious awareness. Initially we point out their behavioral choices serve to reinforce these dynamics, oftentimes leaving them in a constant state of psychological disappointment.

For example, an adult client may constantly find herself commiserating with her mother's seemingly unending unpleasant life. The daughter's offering of any number of qualified solutions is thwarted by her mother's catastrophizing negativity.

The daughter comes away from each phone call feeling exhausted, angry and even guilty. Still she repeats this interchange on a weekly basis. Wanting to end this routine she seeks the therapist's assistance in understanding this dilemma.

An initial social history reveals that she is an adult child of an alcoholic family system. Her early childhood was one of privilege due to her father's successful career as a computer salesman. However, as his drinking increased so did the number of failed jobs and businesses. The enabling mother exhausted herself in supporting her husband despite his alcohol-fueled philandering. This all culminates in her mother going through a series of health and financial crises. The client feels sorry for her mother and tries to fix these situations in a myriad of ways. As usual in these systems, the abuse of alcohol is never posited as a possible cause of these woes.

The beginning phases of therapy find the client quickly realizing, at least intellectually, that her role is fruitless. Armed with the cognitive knowledge of her role in the dynamic the client is chagrined that she continues to engage in the unproductive conversations with her mother. This is when the chess analogy becomes useful.

In learning chess, the neophyte initially learns how each piece moves across the board. Armed with only rudimentary tactics any attempts to play a game with even a novice player are quickly met with disaster. Just like our client, possessing only elementary insight, she is no match for a mother with years of manipulative practice. As one's knowledge of chess progresses, one is able to learn from mistakes, sometimes as the piece is moved. Before taking a finger off the chess piece, the player can sometimes see the opposition's trap. Occasionally we see the trap milliseconds after we pull our hand away, which can be a learning experience in itself. This accumulated experience leads to developing strategies where the chess player can see several moves ahead, rarely making a careless mistake. Finally a chess master (so I'm told) can see several games ahead, quickly surmising his opposition's strategy.

Armed with this analogy our client soon found that she could anticipate when the conversations with her mother would become frustrating and guilt inducing. She learned, in effect, how the pieces moved. Her mother would complain about her miserable life and her daughter would counter with possible solutions only to have her piece taken in the form of a rebuff. Sometimes she caught herself in the midst of a conversation, sometimes minutes after she hung up the phone. Still, knowing where she made the mistake reduced the amount of guilt and frustration. Quickly thereafter, the client was anticipating several calls ahead, beginning to forecast when things would be especially stressful; for example, around holidays or planned family gatherings. In session she could practice counter moves to any attempts of pity inducing behavior. She also developed a deeper insight into her own motivations for fixing her parents situation. It seems she also missed that time when her parents were well off, reporting that this was the happiest time of her life. Her various attempts at helping her parents was an unconscious attempt to magically return to that time, thus avoiding the shame and disappointment associated with growing up in an alcoholic household. This insight greatly reduced the number of times she was caught up in the emotional phone calls.

"Behavioral change is like the formation of an island."

- In this analogy, the therapist compares the practice of new insights, new behaviors, and new ways of thinking to the formation of an island.

- Use this analogy in situations where clients and patients, despite their hard work, remain frustrated with their progress.

- Therapeutic impasse

- Behavioral practice

Clients regularly reach an impasse in treatment. They have dutifully attended every session and diligently worked to uncover unconscious motivations for their maladaptive behavior. They have practiced different coping mechanisms and tried new cognitive patterns; still they essentially remain unsatisfied in their life. This frustration spills over in the therapeutic setting, and they feel as if they are making only minimal progress toward treatment goals. In reaching this impasse, many begin to give up hope. During this frustrating phase of therapy, I often offer the following analogy.

I tell them that sustained behavior change is often like the formation of an island. If one is to look at the ocean's horizon, they see no landmass protruding from the water. However, as we all know, beneath the sea there is a cumulative building up of volcanic rock that after a time finally produces an island. I tell them that though they cannot immediately see sustained behavior change they should not discount the small behavior changes they have executed. The constant practice of these new insights and coping skills are building up one on top of the other, just as an island forms, and soon they often coalesce to form a sustained adjusted new self. I encourage them that soon they will look off into that horizon and see the protruding land mass. Moreover, what will be interesting to them is that it will seem that it has been there all along.

"Diverging lines"

- This analogy demonstrates that progress in psychotherapy, like diverging lines, may seem imperceptible at first, only to grow greater and greater.

- This analogy is used to show our patients that acquired skills practiced repeatedly, lead to exponential and cumulative change.

Similar to using the 'island formation' analogy as a means to motivate the client to continue to pursue their therapeutic gains, the analogy of diverging lines is another way to help the client see that sustained behavior change is a cumulative process that may not be noticed right away. In this scenario, the client experiencing an impasse in treatment may begin to feel hopeless about long-term behavior change. They may report, "Doc I've done everything you've asked and I still don't feel significantly different."

At this juncture I ask them to imagine two parallel lines. I suggest that one line represents their condition coming into therapy, and the other line, not too far apart from the other, represents their perceived static progress. At this point, I challenge their perspective, instead positing that there has probably been a slight degree of derivation in the second line. Together we identify mutual positive movement that has been accomplished toward treatment goals. Most often, they agree to this concept. I then ask them to follow this slight divergence, barely noticeable at first, over a long distance. When they do, they often see that that over the long haul this second or progress line eventually deviates significantly from the line analogous to their premorbid condition.

"Cutting weight or throwing out unnecessary weight in order for the hot air balloon to rise."

- Throwing over excess weight in order to get a hot air balloon to rise is analogous to the patient extricating themselves from destructive relationships and maladaptive patterns in facilitating self-growth.

- This analogy can be used in myriad situations to demonstrate to clients the need to rid themselves of destructive stressors.

- Codependency

Codependent patients struggle with giving up various stressors that weigh them down. Whether it is unproductive relationships, setting faulty boundaries, or their physical health, many patients are reluctant to cut ties with these established patterns. It is usually motivated by fear of some type of reprisal, rejection, or lack of self-esteem. These emotionally abusive relationships, maladaptive eating patterns, and poor boundaries give them some modicum of comfort, pleasure or potential for validation. This fear of not having healthy means of replacing them is very real; still these dead weights are a hopeless burden. Most patients realize that if they were feeling at their psychological best they would have little time for these unproductive elements.

To help patients understand this, I ask them to visualize themselves as a hot air balloon that is in danger of going down if they continue to carry such dead weight. While maintaining sensitivity to patients' deep-rooted adaptive beliefs about these patterns and people, I then ask them to envision them as dead weight that is causing the balloon to sink. I ask them that if necessary, which weights (patterns or relationships) would they cut, ranking them from least necessary to most. This visualized analogy often allows the client enough emotional distance to one, get them to see themselves as not moving forward, and two, have them reevaluate the necessity of their environmental stressors.

"The Analogies of Absorption"

- These analogies compare various articles of absorption (i.e. .facial tissue, paper towels, sponges) to the ego's capacity to handle stressors, regulate affect, and maintain a stable sense of self.

- This analogy is useful for aiding clients to understand relative strengths of their self-makeup.

- Ego strength

- Self psychology

One measure of ego strength that seems to run through all theoretical orientations is the ability to absorb. Whether it is daily stressors, or loved ones projected aggression, the ability to absorb these attacks to the self without significant deterioration or regression is a sign of an integrated and sustained sense of self.

Periodically an individual makes their way into my office worn down by the cumulative absorption of emotionally giving to others with little reciprocation. These super women and men find themselves exhausted for the first time in their lives. They complain that stress they once were able to shrug off is now incapacitating. They become emotional at the slightest provocation, sometimes welling up in tears. They are literally 'leaking out', no longer able to absorb others projections or demands, their reserves depleted. They are frightened of this phenomenon and fear it is a permanent transition, which is sometimes the case. However, more often than not, it is a transient phase requiring nothing more than setting boundaries and rearranging priorities. To help these individuals understand their predicament I offer the analogies of absorption.

Facial Tissue

These individuals have difficulty absorbing even the slightest degree of stress or narcissistic injury. Like trying to wipe up a spill with only a tissue, they saturate quickly, often fragmenting into tiny pieces after prolonged use. These are akin to some of the characterological people we see in our practices (borderlines, dependent personalities, or the abandoned narcissist). These individuals constantly find themselves in a degree of decompensation or fragmentation trying to deal with the many emotional 'spills' of life. They cope by avoiding these messes altogether, or sometimes linking themselves with others that manifest greater absorption capabilities.

The Paper Towel

These individuals, while possessing greater absorbency than those patients whose ego functioning was compared to facial tissue, are still very limited in the amount they can deal with. They too, often link themselves with others who are enabling caretakers. They can usually take care of themselves and perhaps even a job or a lone child, before they risk decompensation, but can handle little else. These individuals are often the slovenly couch potato husbands or the minimal mother clothed in her robe past noon.

The Everyday Store Bought Sponge

This category comprises most of us. Individuals with stable object-relations, adequate ego strength and equipped with an effective array of coping skills. We absorb most emotional stressors; our toddlers temper tantrums of projected aggression, our jobs and spouses, and any unforeseen 'spill' that presents itself. We can usually begin to feel saturated after a time and notice the need to 'wring ourselves out' in the form of a well deserved vacation (perhaps 'drying in the sun' to continue with the metaphorical language), pawning off the children for a date night, a warm bath, or by taking a 'mental health day' on the golf course or at the ball game.

Once or twice during our lifetime, these types of stress relievers simply are not sufficient and the situation culminates in total saturation. Because we are usually aware of our limitations, we are likely to ask for personal or professional help when these situations arise. Once dried out we are again ready to tackle those upsets that come our way.

The Super Sea Sponge

These somewhat rare individuals' absorption rates are akin to the giant sea sponges seen on nature programs or accoutrements at posh day spas. These giant sponges are able to absorb multiple large spills without even the slightest degree of saturation. Individuals who fit this category are the superwomen and men we all simultaneously envy and admire should we run across one. These seemingly impossible individuals appear to 'do it all'; not only do they take care of themselves, their families, and their careers, they also are our church elders, little league coaches and homeroom mothers. Because they do not always 'see it coming', they are prone to fall hard when some cumulative or exorbitant stressor finds them. Because of their super absorbency makeup, they are susceptible to becoming supersaturated. When they go to wipe up a usually absorbable spill, they now, because of their supersaturated state, leave more moisture than they pick up. Inexperienced with this feeling they continue their old coping skills finding themselves increasingly frustrated and frightened that they can no longer absorb and are leaking out emotionally.

"The poisoned well"

- Continuing to seek validation from a failed relationship is compared to drinking from a once fresh but now toxic water source.

- This analogy is useful for individuals having difficulty moving on from an ended relationship.

- Abandonment anxiety

- Affect regulation

- Relational breakup

A horrific break-up is often the impetus for one to enter into therapy. The desperate anxiety of abandonment is unbearable, overwhelming existing coping mechanisms in even the most stable of individuals. Unfortunately, break-ups are rarely clean and mutual. One or both parties linger on in hopes of recapturing the magic that once was. Stable self-representations go out the window, and the individual capitulates to seek self-esteem nourishing boosts from the one individual who can no longer provide it. The prospective client is frustrated and self-punitive. They continue to seek out that individual when desperate, hoping for validation only to go away feeling even more depleted than before they called. The initial relief of hearing the others voice temporarily staves off their anxiety.

Soon the conversation deteriorates into staid and maladaptive communication patterns ending in why the relationship had to terminate in the first place. This self-defeating cycle repeats, and the reserves of self-esteem deplete even further. Before one realizes it, they are experiencing the lowest self-worth of their life. Friends and family are of little solace, and it seems that the only person who can restore their self-esteem is the lost love.

These individuals come to treatment for reasons twofold: one, to permanently deal with their abandonment anxiety, and two, to understand "Why do I keep calling him/her?" In some cases it is, "Why is it when I finally feel as if I can pull away, she calls me and drags me down?"

In many instances, the relationship was the best of their life, often following the end of a particularly dysfunctional or unsatisfactory pairing. Forged on the rebound, the relationship was magical, and the client cannot comprehend how it soured or why it can't return to its original state. Cognitively and intellectually, they know they must forge on, but to do so means the loss of the ideal. In their depleted state of self-worth, they are afraid of moving ahead, unable to realistically see themselves in any other relationship.

In these cases I often offer the analogy of the poisoned well as a means to assist them in understanding the requirements of their recovery. I describe a well, but not any ordinary watering hole. This well had the best water ever, clean, fresh, and satisfying. It was a wonderful find and never failed to satisfy. For some reason, the well became toxic and unhealthy. There was no real explanation for the well turning toxic, so the individual continued their daily treks for water believing that it was only a temporary problem. They knew of no other well that could quench their thirst in as fine a way. They kept going back, and although the water looked similar, it continued to contain a degree of toxicity. They did not realize they had become so dependent on the well, and although they knew or had heard of other wells, they feared the distance they would have to travel. Besides, what if the well changed while they were traveling to another? They continued going back to the poisoned well but each time they became sick.

They finally tired of the deleterious effects from the well and reluctantly forged on to find other sources in which to quench their thirst. Along the way, and sometimes after a great distance, they did find other watering holes. As expected they were not as fresh tasting, or satisfying in the same way, however they provided enough water to survive and they were never toxic. Eventually these individuals found enough wells of good quality that although they missed the original they were no longer tempted by the poison well.

"Be aware of the body blow in boxing."

- The collective process of absorbing others projective identifications is analogous to the exhausting and cumulative sustaining of body blows throughout a boxing match.

- This analogy is helpful for those clients forced to deal with individuals with characterological traits.

- Projective identification

- Borderline personality disorder

In boxing, it is a well-known strategy to throw punches to the midsection during the initial rounds of the match. While no single blow will knock a person out, cumulatively they take their toll. This succession of body blows during the earlier rounds results in the boxer experiencing labored breathing during the later rounds. Fatigued, he drops his arms to protect his flank. With the arms lowered, the boxer is now susceptible to the one knockout blow to the head.

These body punches are insidious in that the boxer does not really feel each individual punch. Only after they have reached a cumulative threshold do they then serve their tactical purpose.

Much is the same within inter and intrapersonal psychological functioning. Specifically, the psychoanalytic phenomenon known as projective identification acts much like the body blow in boxing. Projective identification is essentially taking on or absorbing those defensive projections that are thrust upon us in our close interpersonal relationships.

Projection is the classical defense where an individual dispels those perceived negative characteristics of the self by projecting them onto others via hostile interchanges. For example, if a person is overwhelmed by their own feelings of insecurity they may choose to constantly accuse their spouse, parent or employee of that same trait. Because there is usually some small kernel of truth to these accusations, the receiver of these messages begins, after frequent exposure, to expand this trait sometimes making this a dominant characteristic of the self.

It is my belief that we see many of these victims in individual and conjoint psychotherapy as members of an embattled and insidiously abusive relationship. These individuals are usually of initially strong psychological make-up, but have found themselves hooked up with an individual possessing many characterological traits. Because of their ample ego strength, they can initially fend off these projections explaining them away with simple rationalizations such as, "She really doesn't mean it. She just gets moody sometimes" or, "She was cheated on in her last relationship, so she has a right to be hyper-vigilant." Just like the boxer, this person is unaware of the cumulative strength of projective identifications. However, after frequent exposure, the recipient becomes vulnerable to taking on these punitive interjects, and soon they experience plummeting self-worth.

"Air is to rational thought as fuel is to emotion: You need the correct mix for good communication."

- The right balance of cognition and emotion needed for effective communication is similar to the need for the right mix of fuel and air needed for combustion engines.

- This analogy is useful for assertiveness training boundary setting, and anger management.

- Assertiveness skills

- Communication skills

- Anger management

This is a multipurpose analogy that has demonstrated utility when teaching assertive skills, helping patients to manage explosive anger, or aiding more passive patients to set appropriate boundaries by giving voice to their emotions.

In using this analogy, I first offer a rudimentary explanation of how a combustion engine works (a rudimentary explanation is all I can offer). Before the advent of modern fuel injection, in order to make an engine run there needed to be the proper mixture of both fuel and air. This process all takes place within the carburetor. As most of us have experienced getting to our car 15 minutes late for work, if there is too much air and not enough fuel the engine simply does not turn over. If there is too much fuel and not enough air in the mix, the carburetor floods and the engine, failing to ignite, is accompanied by the unmistakable aroma of gas.

I propose that effective communication also requires a proper mix of emotion and rational thought. Excess emotion without cognitive mediation emotionally 'floods' the interaction. This results in the messenger coming across as overbearing or even threatening. Too much rational thought without the corresponding degree of emotion often renders the message impotent or ineffectual.

"Climbing a mountain one step at a time"

- Progress in psychotherapy is akin to climbing a mountain.

- This analogy is used to aid clients in setting realistic short-term goals along the way to longer-term sustained changes in self. It is also useful in helping patients gauge already gained improvement.

- Goal setting

- Treatment planning

Perhaps the greatest advantage we have as therapists is seeing our clients over an extended period of time allowing us a greater viewpoint on their progress, a perspective that oftentimes is outside of their awareness. Much the same way as we cannot always accurately measure the physical growth of our children, needing a periodically visiting relative to point out the obvious; a patient cannot accurately gauge their progress during therapy.

During periods of therapeutic impasse or plateaus, patients and clients often complain that they cannot see progress. For these clients, comparing climbing a mountain to progress in therapy is helpful. Comparing therapy to a journey? Now there is a novel idea. However, just like an old wife's tale or farmer's wisdom, we cannot completely ignore their heuristic value.

Sharing with the client aspects of hiking as a means of looking at therapeutic progress is indeed useful. For example, during a particularly difficult ascent, one often only looks ahead at the seemingly insurmountable peak, without taking time to occasionally stop and look back to where they have come from. During this comparison, I recount for the patient the many treatment goals and sustained behavioral changes they have already attained.

This analogy can also be helpful during the beginning stages in treatment or when individuals get discouraged about the difficulty that remains. Telling them to just focus on the immediate steps ahead (baby steps anyone?) without looking too far up the trail is often encouraging. Then they can be assured that once started, they can occasionally stop and look back, and they will be pleasantly surprised at the distance traveled.

"The farmer and his crops"

- This analogy humorously compares a patient's resistance to a farmer pessimistically forecasting his crops shortly after seeding.

- This analogy is used to join rather than confront a client's resistance or impatience.

- Joining

- Patient resistance

- Reality defense

An overly pessimistic and resistant client can be frustrating to even the most experienced clinician. Of course a healthy dose of pessimism and impatience is a characteristic of good ego strength, but it can also be an ingrained defense against failing to meet expectations or confronting old schemas. Therapeutic interventions geared to break through this type of resistance are often futile. Sometimes, and I repeat sometimes, analogies and other transitional languages (similes and metaphors) can be of aid. One that I lifted from a client, with permission of course, involves the ever-pessimistic Midwestern farmer. I preface this allegory with what my patient told me; that a highly evolved since of dread is a prerequisite for being a successful farmer. Without it, he explained, one could not demonstrate the perseverance and diligence needed to be successful. By therapeutically joining with the client's pessimistic resistance versus confronting it, one may to some degree avoid the 'digging in' effect.

My patient recalled a story of an uncle who yearly planted his crops. Invariably the next day he would go out to the overturned field and kneel down, rubbing his chin while periodically scratching at the dirt. After staring at the day-old seeds for a couple of minutes, he would reply in all seriousness, "They're not going to make it." Of course my patient, an old farmer himself, ended the tale stating that save one or two years, this uncle always produced bumper crops.

"Implementing change is like a bear learning to fish: it requires practice and patience."

- The implementation of new behaviors gleaned in psychotherapy is analogous to the learning curve of a young bear learning to fish.

- This analogy is useful in helping clients maintain patience in practicing out of session behavior.

- Learned trials

- Reinforcement

Clients beginning to implement new behaviors gleaned from therapeutic insights can become so frustrated that they threaten to abandon the whole process and return to their previous comfortable yet maladaptive strategies. When this comes up in therapy I offer the analogy of the bear fishing in the river. When we watch these amazing behemoths gracefully scoop the fish in seemingly impossible fashion from the rushing rapids, we marvel at this talent, as it must surely be innate. I share with them a nature program I once viewed showing a much younger bear awkwardly stabbing at the elusive fish producing only large splashes while occasionally slipping from his perch and into the river. Most patients have seen such a program or easily grasp the visual. I implore them to forgive their awkward first tries and stick with their plan for at least two more weeks and then realistically review their progress. As expected, at the end of this trial period, most do not claim expertise but they no longer feel like the young bear in their initial stabs toward new and long-lasting behavior change.

"Peroxide in the wound"

- Cleansing a wound with peroxide is similar to the process of uncovering forms of psychotherapy

- This analogy is helpful in fighting patient resistance explaining the need to rework sometimes-painful early events.

- Uncovering psychotherapies

- Resistance

I do not feel it is too much of a leap to compare the process of psychotherapy to the cleansing of a wound. The internal dynamics of a struggling individual are much like a festering wound, one that has been inadequately treated with diluted over-the-counter medication or ineffective home remedies that only partially heal the injury. Worse yet many individuals simply numb their pain through self-medicating means such as alcohol or drugs. During the beginning phases, individuals are resistant to digging beneath the surface and reopening old wounds. Many types of psychotherapy, most predominantly psychodynamic methods, require the recalling of previously painful events. To allay a patient's fear, I compare psychotherapy to pouring peroxide into an infected wound; it initially hurts more, but the cleansing properties are long lasting and effective.

"The Frying Pan of the Plateau as opposed to the Frying Pan of Descent"

- This analogy compares the precipice of a plateau to the inevitable, difficult decisions our patient's and clients have to make.

- This analogy is helpful to clients in an urgent situation that demands impending and sometimes drastic action.

- Radical acceptance

We often find our clients in the throes of having to make very difficult decisions. This choice, containing no real tangible sense of instant gratification, often brings the anxiety of the unknown. Week after week, we observe clients conjuring up defenses to avoid the inevitable. Classic examples of these dilemmas include; unwanted pregnancies, emancipation and moving on from unforeseen losses. In order to conceptualize and encompass their situation as a whole, I might relate the following analogy that represents their current dilemma:

Your situation is analogous to someone on the run who has found themselves at the precipice of a steep canyon. Treacherous but negotiable, it seems to be a perilous journey. However, if you look back from whence you came, you realize that the force that you most certainly need to get away from is closing at a sure but steady pace, like the position Paul Newman and Robert Redford's characters found themselves in the movie *Butch Cassidy and the Sundance Kid*. You find yourself looking back down the canyon, and then looking back at the closing gap of your enemies. Sometimes you find yourself sitting and doing nothing hoping by some miracle your predicament will change.

You sense your urgent need to negotiate the canyon, but your reluctance to choose this inevitable course is rooted in the canyon's danger and sense of the unknown. It is important to focus on this option, not panicked by but aware of the closing gap. By not acting now, the situation will only become more imperative, perhaps leading to panic and an impulsive decision. Perhaps your avoidance will simply lead you to succumb to your pursuers, resulting assuredly in self-hatred because you did not at least try. I then ask the client to metaphorically imagine their descent down the canyon in measured yet optimistically hopeful steps utilizing realistic coping mechanisms that they already possess.

Analogies Using Classical Theories

A majority of this book has been a transition away from the use of theoretical language, and for good reason. However there is richness to the therapeutic heuristics contained in the plethora of classical experiments used to prove these theories. I'm speaking here of those classical experiments, such as: Skinner's reinforcement schedules; the learned helplessness experiments Seligman conducted using the dog and the electrified grid; and even Alfred and the rat. The cogent findings of these studies are so direct that they are easily translated into practical situations our clients face every day. I have lost count of how often these standards of Psychology 101 have spontaneously come to me during session to benefit clients in a myriad of therapeutic situations.

"Skinner's fixed and variable ratios of reinforcement"

- This analogy compares removing oneself from a continuously destructive situation to Skinner's operant conditioning concept of fixed and variable schedules of reinforcement.

- This analogy is helpful in explaining to patients how they inadvertently reinforce others abusive behavior in a myriad of situations.

- Extinguishing negative behavior

- Operant conditioning

- Boundary setting

- Fixed ratio

- Variable ratio

We all remember the classic bar pressing-for-food studies utilized to explain Skinner's theory of operant conditioning. In these studies, a rat is trained to press a bar a fixed amount of times in order to receive a pellet of food. To extinguish this behavior, one simply quits dropping a pellet of food once the rat registers the correct and fixed amount of bar presses. As we recall initially the rat's bar pressing increases as he attempts to obtain the food pellet. If the rat is consistently rebuffed the bar pressing tapers off quickly and the reinforced behavior of bar pressing is indeed extinguished quite quickly. However, if during this period of extinction, where the rat's bar pressing behavior is on the increase, one intermittently drops in a food pellet, the rat's bar pressing behavior now rapidly increases. Now he has been placed on a variable ratio schedule where this behavior is reinforced after an unknown number of bar presses.

This study works as an analogy in any number of clinical cases. For example, in many divorce and custody cases one party often bombards the other with abusive calls ostensibly posed as 'regarding the child'. I often meet with the exhausted client who, while trying to desperately move on, continues to reinforce the other's hostility by taking these calls that contain hidden agendas.

This study also works as a useful analogy for individuals trying to break off a relationship, or for codependent parents who are trying, without much success, to emancipate from their emotionally and financially draining adult children. It always amazes me that people usually quite quickly comprehend the rat's behavior.

"Skinner reinforcement and the child tantrum"

- Bribing a child to gain compliance is analogous to Skinner's behaviorism concept of Negative Reinforcement.

- This analogy is helpful for patients dealing with progressively misbehaved children.

- Operant conditioning

Another behaviorism concept, negative reinforcement, posits that we will perform some behavior in order to get another behavior or stimulus to stop. The classic example involves exposing cats to a noxious sound. Shaping through successive approximation quickly takes place, with the cat quickly learning to perform another behavior to get the noise to stop. A practical every day example of this concept is taking an aspirin to get a headache to cease.

How I use this analogy in practice most often involves a child-parent interaction pattern where the parent gives in to a child's tantrum or disruptive behavior in order to get the child to temporarily cease the 'bratty' behavior. I explain to the exasperated parent that the child now has them conditioned; the child's negative behavior becomes the noxious and aversive stimulus that can be stopped only by the parent performing some 'giving in' behavior (i.e. buying the begged for toy, taking them to McDonalds if they quit crying, etc.). Usually the parents intellectually understand this concept but continue giving in complaining that previous attempts at alternative discipline only made the behavior worse.

I explain that this escalation of the temper tantrum is expected as the child now ups the ante to see if you will eventually perform the past behavior of capitulating to get them to stop. This is a natural process in the transition of regaining control. The behavior will likely increase until they finally (and this may take dozens of attempts) realize that this behavior will not get the desired response.

"Approach – Avoidance and the
Extra-marital triangle."

- This analogy compares the concept of approach-avoidance to the partial fulfillment of staying in an extra-marital affair.

- Extra-marital affairs

- Part-object introjects

- Freudian concept of Tension reduction

The approach-avoidance concept posits that many phenomena require maintenance of equilibrium. Our own relationship with the sun is a good example, if we get too close, we burn up, and if we drift too far, we freeze. Our patient in the throes of an extra-marital affair, whether they are the married individual or the paramour, need to understand this concept so as to better understand their options.

Likely, the person cheating is unhappy with some aspect of their relationship. They have found in the affair at least a partial and transient fulfillment of this unhappiness. However, for any multitude of reasons (fear of loss, children, financial) they demonstrate difficulty leaving the marital relationship entirely . They in essence want to have their cake and eat it too by keeping a comfortable distance between the marriage and the affair. They are likely to make false promises to leave their spouse to be with their lover only to recoil, as they get closer to fulfilling the improbable pledge. They remember the impending consequences of loss if they are to follow through. As they distance themselves from the affair, they may lament the current state of their marriage and seek to bring the affair closer.

These clients and patients are sometimes stuck in these patterns for years with only the promise of partial fulfillment possible. It is important for the patient, wherever they fit in this triangle, to understand this maladaptive pattern and gain insight and responsibility for its continuance.

"Seligman's Learned Helplessness: The dog and the electrified grid experiment."

- This analogy demonstrates the utility of Seligman's classic experiments to every day clinical situations.

- Learned helplessness

- Authentic happiness

- External-internal locus of control

- Depression

The classic theory of learned helplessness and its implications in depression are obvious and well grounded in scientific literature. In summary, if one does not perceive control over their environment, they will, over time, concede to a life of victimization and apathy. Various cognitive schemas result from this condition including: "What is the use? I can't do anything about it anyway" or "I try to do it but something always gets in my way." These schemas are hard to break and many times have a strong basis in reality. These patients have usually experienced a history of trauma or abuse where the will of another so significantly impinged on them that they had no real recourse other than to be helpless. To break through this reality defense, it is paramount that the client gains at least some perceived sense of control over their environment. To illustrate this phenomenon in an encapsulated and somewhat emotionally distant way, I relay the classic experiment involving the dog and the electrified grid.

In this classic experiment, a dog is placed on a grid that is divided in half by a very short fence. The experimenter can electrify either side of the grid. The dog is conditioned to the sound of a buzzer, notifying the animal which side will soon be electrified. The dog was quickly conditioned so that when the buzzer sounded, he would have to jump the fence to the other side to escape being shocked. As long as the animal had perceived control over when the grid would be electrified, he maintained relatively unaffected. It was only when the experimenters manipulated the buzzer in various ways that the dog began to act in erratic yet wholly understandable ways. When they changed the interval between the sound and the time when the grid would be electrified, the animal showed signs of anxiety but actively attempted to try to maintain control of the environment. Further manipulations of this poor canine's environment (thankfully these experiments are no longer sanctioned) caused continued attempts to make sense of his world. In addition to the interval between sound and electrocution of the grid being changed, the experimenter would sound the buzzer and randomly electrify one or the other side of the grid, further limiting the animal's sense of perceived control.

Finally, the experimenter would electrify both sides of the grid simultaneously and the dog did the inevitable. Usually when I relate this simile to the patient, I ask them, "What do you think the dog did?" Sometimes I get out of the ordinary answers such as "He bit the experimenter", which is probably more wish fulfillment than an actual guess. However, like the above rat study, the patients more often than not know the right answer. Bearing the lifetime of like experiences, they answer in a resigned tone, "He simply gave up and lay there." The way in which many patients answer this question indicates a profound emphasis of understanding. Comparing this analogy to one's own experiences, it is often an "A ha!" revelation. Coming at their faulty beliefs from this angle is a great example of why this type of transitional or metaphorical language has such great therapeutic value.

It now allows the patient the opportunity to focus on changing maladaptive automatic thoughts and behaviors. This learned optimism, as Seligman coined it, helps the patient change their perspective by focusing on what aspects of their environment they can control.

"Analogy of the diminishing return curve"

- This analogy compares the physics concept of a diminishing return curve to show how individual's previously productive, yet maladaptive coping skills, self-perceptions and automatic thoughts breakdown under stress.

- This analogy is useful in helping clients expand their repertoire of productive coping skills.

- Ego ideal

- Automatic thoughts

- Type A personality

The diminishing return curve, a physics concept, essentially posits that something used to produce results, if used too often can eventually lead to reduced production. For example working out with weights does indeed increase muscle mass and furthermore, the more one works out the more muscle mass that is gained. Essentially, someone who works out 3 times a week will gain more muscle mass in a faster time than someone who works out 2 times a month. However, and thus the term diminishing return curve, if one works out 4 times a day they will begin to see diminishing results due to the muscles becoming fatigued and not having time to rest.

This has been an extremely useful analogy in helping some of our high achieving and perfectionist patients to see that their automatic thoughts and hard driving behaviors can have limitations. In fact these tried and true means of achieving results can, under periods of extreme stress or bouts of physiological depression, actually produce diminishing returns.

A good example of this phenomenon is the patient who uses self-negative ideation to push himself to achieve. Perhaps a product of a harsh ego ideal, the patient cuts himself down at the slightest hint of failure, pushing himself even harder to produce results. This has been adaptive in the past and many times this type of motivation has indeed produced achievement. Conceivably this is an attempt to prove to some internalized or real parent that they are not a loser or "stupid." However, under exacerbating stressors, this self-negative ideation does not provide the desired results.

The patient becomes increasingly harder on himself or herself, only to see diminishing outcome. The analogy of the diminishing return curve is helpful at this point to subtly validate their existing coping skills, yet have them see the ultimate flaw. It has been useful to help the client expand their repertoire of coping skills to achieve the same results.

"The physics wave"

- The diminishing impact of a physics wave is analogous to the lessening of a psychological trauma over time.

- This analogy is useful during the crisis phase of psychotherapy.

- PTSD

- Crisis intervention

Patients frequently initiate therapy as a response to a devastating loss. Their usual capacities to deal with such a calamity have been depleted and maximized. The overwhelming hurt is too much to bear and they enter therapy looking to reestablish some sort of return to premorbid functioning or equilibrium. Their resources taxed, they cannot imagine how the magnitude of this psychological injury can be diminished. In this state, they look to the therapist for some sense of realistic reassurance that this indeed can be accomplished.

Using the analogy of psychological equilibrium as similar to a physics wave, the therapist can provide relief during the initial phase of crisis psychotherapy. The concept of a physics wave posits that the initial impact of an action creates the largest change and then diminishes over time. Similar to a rock dropped in a pond, the initial impact has rippling effects, but by nature these effects diminish as the ripples travel further from the spot of the initial impact. We can assure our patients that psychological trauma has similar effects; there are lasting consequences or ripples following the initial impact, but these too have a tendency to diminish over time.

Family of Origin Analogies

"Waiting to be pitched to"

- The adult child, stuck in perceptual adolescence, waiting for his parents attention is analogous to a child waiting to be pitched too after everyone has gone home.

- This analogy is useful for clients who are stuck in an enmeshed and placatory role with their parents to the detriment of their own autonomy.

- Identity formation

- Enmeshment

- Necessary loss

- Radical acceptance

- Separation-individuation

- Birth order

There is some debate concerning the scientific validity and theoretical efficacy of birth order, but its presence as a contributing variable cannot be denied. Depending on where a child is born into the developmental life of a family, there appear to be definite and long lasting psychological ramifications. While there is surely a static degree of family dynamics, I also believe they can be fluid depending on several variables. Throughout the years I have been exposed to a group I call forgotten children.

These children usually come along later in the developmental life of the family. Born into larger families, frequently there is at least a 5-year spread between themselves and their next oldest sibling. Characteristic of these families, an older child may have demanded extensive parental attention. They may have been externalizers, frequently acting out in response to a troubled marriage, but nonetheless they often depleted family resources.

Our patient, as an adaptive response to this dominant sibling, forms a passive personality, appeasing the family by obediently listening and performing the required tasks to minimize familial conflict. There is an unconscious internalized belief that if they bide their time, the parental attention will eventually come their way, their patience being duly rewarded. However, this unfortunately never materializes.

Significantly, life-altering circumstances (health, finances, a drunken parent becoming sober, or even divorce) burn out these often older than average parents, leaving them exhausted. Our patients, consciously unaware of their wish to be number one, sometimes remain developmentally fixated in suspended adolescence. They wait for parental direction that never comes. They remain connected to the family dynamics, perhaps continuing in their role as emotional confidant, or good son/daughter, patiently listening to their parent's troubles at the expense of their own psychological emancipation or individuation.

These patients often enter treatment depressed and dissatisfied, perhaps stuck in an endless string of 'McJobs' they are over qualified for, or undecided regarding their college major or graduate school well into their 20's. Their session content is equally divided between complaints about their current situation and commiserating about their enmeshed communication with their parent. A life of passive placating has made it difficult to get righteously angry.

This situation unfortunately describes what Judith Viorst termed a necessary loss. The patient needs to resolve, for the sake of his continued autonomous development, that his unconscious wish of finally being recognized as the important one may never arise. The analogy I use to summarize the patients predicament is that of a youth participating in a sandlot baseball game where they are batting ninth. They wait patiently for their turn at bat.

However, the game runs into people's dinnertime and the game abruptly ceases before our patient has his chance at the plate. Ever patient and obedient, our client waits for them to return after dinner, but nobody ever shows. The child, bat in hand as dusk becomes night, is now illuminated only by the streetlight. The stark recognition of this situation often elicits feelings of sadness and finally anger, as they realize their role in the family may not ever transform from anything but the low maintenance and placating son or daughter.

If this analogy resonates with a particular patient, we use it as a means to establish individuation. Perhaps we can get them to see the benefits of moving ahead to a new baseball game. While recognizing that it can never be as satisfying as playing in that other game, it is realistic that this new game is more under their control and over the long haul, infinitely more satisfying. We compare the pitcher in the game in which they were waiting to parental recognition and direction. Therefore, one needs to search for a different pitcher in which to glean advice and direction. Perhaps an internalized paternal figure or mentor can suffice in the absence of the 'real thing'. Usually we can gauge therapeutic efficacy when the client feels they are 'participating' in their own new game and has satisfactorily resolved that they might never play in the original one.

"The mother and the reaction to mud pies analogy (3)"

- These analogies compare the quality of early interactions in the mother-child dyad and subsequent self-representations to a parent's reaction to a child's first mud pie.

- These analogies are helpful in explaining to clients the importance of mirroring and internalized representations of self.

- Cognitive schemas

- Object relations

- Mirroring

- Internalized voice

- Self-esteem regulation

Whether it is the development of sustained cognitive schemas, or the stable self-representations posited by object relations theorists, most theories of self-worth and self-esteem are heavily grounded in the early parent-child dyad. It is often said that early mirroring experiences between a parent and the child determines an individual's ability to regulate and maintain a stable sense of self-worth and self-esteem. In an attempt to gauge early parent-child relationships during the initial phase of treatment, I rely on this analogy as a means to understand the quality of these relationships free of theoretical jargon:

A patient usually hints at the quality of these relationships; however some individuals need this dynamic illustrated more clearly. I often make this analogous to how a parent might respond to their child bringing in a just completed mud pie. In an ideal parenting situation, the optimum parent would respond in a very excited and proud way such as "Wow, look what you made!" or "That's the best mud pie I have ever seen". Despite the parents noticing the potential mess, the good parent would instantaneously decipher their child's obvious pleasure in their accomplishment and 'mirror' back to them in kind. If enough of these critical experiences are positively mirrored back to the child, proper affective development can occur and a stable self-representation of an autonomous and industrious child can emerge.

I then offer the client less adequate parental responses. Does the parent respond to the child's creation by neglect? Perhaps they are too busy fighting with their spouse. Maybe they are depressed on the couch, or sleeping off a hangover from the previous night's substance abuse.   These cumulative experiences often lead to an adult being emotionally bereft or depressed themselves.

Another parental response to the child's creation might be one of critique, such as, "That's not quite right honey. A good mud pie would have the edges cut off better." Or maybe a parent would be overly concerned about cleanliness and make the child feel unduly guilty, missing altogether the child's affective experience. If exposed to too many of these exchanges, the child may subjugate their own affective experiences, instead struggling to mirror what their parents are feeling. This sometimes leads to the ineffectual adult, unsure of their own feelings making them prone to others demands. They will often base too much of their self-esteem on pleasing others. A good example of this is contained in Alice B. Miller seminal work, *The Drama of the Gifted Child*. Simplistic, it is nonetheless a helpful way of getting patients to describe the quality of their early family relationships. Many times the patient will spontaneously come up with their own hybrid parental response that more accurately describes his or her own unique experience.

The Wizard of Oz: "Pay no attention to that man behind the curtain."

- The inadequate individual that hides behind narcissistic bravado is akin to the man behind the curtain masquerading as the Great Wizard.

- Uses include working with individuals still controlled by a bullying and intimidating parent.

- Abusive parent

- Identifying with the aggressor

- Victims of Abuse

A family member, utilizing methods of guilt, shame and even terror can loom so large that they can affect a family for generations. Decedents of these figures become so paralyzed in their ability to make decisions that they still fear their retributions even if they are thousands of miles away. A necessary component in working with the victims of these circumstances is to convince them that although at one time they needed to cower down as a means to survive, these individuals presently do not have as much power as they think. It is necessary to convince the patient that most often the perceived towering figure wielded this manipulative power as a defense against his or her own inadequacy. An analogous example I often use to illustrate that these individuals bark is much bigger than their bite follows.

I ask them to remember the scene in the all-time classic *The Wizard of Oz* where the sundry characters first confront the Wizard. His ominous presence is characterized by the menacing giant character on the screen with the booming voice, complete with smoke and flashing lights. The friends are terrified initially and it is not until Toto slides the curtain open, revealing a nebbish old fellow furiously working a large wheel to conduct the elaborate presentation, that they realize who the wizard actually is. Who can forget the classic line as the traveling mates begin to catch on to the chicanery: "Pay no attention to the man behind the curtain!" I hope to get the client to realize the analogy between the fearsome wizard of their youth, and the more realistic and often pathetically inadequate character underneath the facade. This sometimes allows the patient to begin to understand their irrational fear in relation to the perceived power of this individual.

"Refraining from rocking the boat in calm waters"

- An individual's need to create chaos out of habit is compared to rocking the boat in calm waters.

- The benefits of this analogy are useful in working with clients who, used to growing up in a chaotic environment, often create havoc in their own life because calmness elicits anxiety.

- Cognitive schemas

A majority of my caseload are men and women who grew up in profoundly chaotic households. From these experiences arose coping mechanisms designed to at least temporarily stem the chaos. However, these adaptive means were only partially successful and because of this, most of these individuals are left with a residual feeling of constant low-grade anxiety. I am thinking here of the patients who endured the most chaotic of environments including growing up with a raging alcoholic, victimization through violence, or even sexual abuse. Most of their lives were spent in continuous and intense emotional fear. In a sense, they were waiting for 'the other shoe to drop'. Many of these individuals, unable to tolerate the unrelenting anxiety, unconsciously and repetitiously create their own chaotic life situations so that they can once again practice the tried and true yet maladaptive coping mechanisms of their youth. Over time, the psychologically motivated patient has painstakingly broken the pattern of created turmoil, and through the grace of God or hard work, stumbled on a healthy relationship. As this healthy relationship continues, there is anxiety due to waiting for a frenzied situation to develop. When this expected chaos fails to materialize, these patients are often tempted to act on their predilection for seeking maladaptive chaos, and may have already begun the process when they enter your office. Many times they are men and women who express dissatisfaction with their marriage and find themselves attracted to less savory partners. To aid the patient in seeing the connection between self-creating chaos as a means to reduce the anxiety, I offer the following analogy:

I compare it to a canoe journey, one that has been consistently fraught with dangerous rapids, and that most of their rowing skills have been predicated on negotiating such rough waters. I tell them they have worked so hard to escape these torrents that they are unsure of what to do when encountering such calm waters. They are so used to expecting the next torrent, that they become increasingly anxious in anticipation. In order to reduce the anxiety, they may purposely rock the boat because then they at least know what to do to steady it. I assure them that over time they will become used to the calm waters and develop new skills in which to tolerate and then enjoy the diminishing anxiety.

Grabbing the brass ring from the merry-go-round"

- Grabbing the brass ring of a merry-go-round and still feeling unfulfilled is akin to coming to the realization that one has lived the lives others have wished for them.

- Uses for this analogy include; working with the perfectionist patient.

- Identity foreclosure

- Authentic self

- "Drama of the Gifted Child"

- Parental narcissism

- Over-controlling parents

A number of extremely high functioning patients enter therapy amidst an existential dilemma. Seemingly possessing the perfect life (good job, upwardly mobile status, burgeoning family, and good health) they still complain of a persistent dissatisfaction. They feel their lives are not their own. They are confused. They state they have done everything they were supposed to do yet still they maintain a disaffected feeling of emptiness.

The phrase "everything I was supposed to do" usually triggers further exploration of early child-parent relationships. Invariably upon exploration, these young adults describe a family where a premium was set on following family expectations. From a very early age, any autonomous thought or action that varied from parental expectation was thwarted and, met with rejection and shame. The child, in attempt to gain parental satisfaction, acquiesces to their demands sublimating their own internal wishes and behaviors. This seemingly adaptive strategy works in the interim, resulting in an obedient child who often excels as they follow the family plan. This comes at a price, however, as the individual ignores their true authentic self. This is in essence the explanation of the how the ego ideal internalizes rigid parental explanations.

As mentioned in a previous analogy it is also the thesis of Alice Miller's classic <u>Drama of the Gifted Child.</u> This sublimation of authentic self lasts through adolescence in what is termed identity foreclosure. Essentially, it posits that during identity formation in adolescence this individual discards the usual rapprochement issues of rejecting early parental messages to experiment with various identities, instead maintaining the rigid identity chosen by their parents. It is hypothesized that these adolescents might later have difficulty when they realize that their perceived identity was not their own.

This is precisely what I have found in these individuals coming to therapy. They have met all of their ostensibly formulated expectations, yet they remain essentially unhappy. They have seemingly come to an unconscious resolution that the pleasing of others, no matter how much recognition is received, is essentially a hollow victory. They are unsure of the etiology of this concept, but nonetheless feel a frightening void in their lives. It becomes paramount in treatment to get these individuals to realize that they have been living to some extent, an unauthentic life significantly chosen by someone else. The goal becomes aiding the individual to understand the reason they have come to such an existential dilemma, thus liberating them to make their own unique and authentic choices in life.

I have found that comparing their situation to reaching for the brass ring on an old fashioned carousel can, for the first time, illuminate and explain their current level of dissatisfaction. As one recalls, on older merry-go-rounds a brass ring was placed on a stationary pole just outside of the riders reach. With each rotation of the carousel the excited children would lean, straining to grab the ring. Most of us would make a couple attempts to realize the futility and return to the spontaneous joy of pretending we were cowboys or princesses. However there were those more motivated and industrious children that would earnestly try each time leaning perilously farther and farther with each attempt. Their singular focus often seemed to negate the obvious joy of the moment. In illustrating this analogy to the patient I suggest they imagine the child actually grabbing the brass ring. Surely this would result in immediate exaltation of a job well done. Of course there would be instant recognition, acknowledgement, and a sense of hard-earned accomplishment. However, shortly thereafter a child would simply be left with only a brass ring, hardly a useful or fulfilling possession for a ten year old. At this point the client begins to glean the comparison to their current dilemma. I often encourage the client to refrain from undervaluing their sense of industry or concomitant accomplishments, stating this is an integral part of who they are. Instead I promote they use these well developed coping mechanisms in making more authentic choices in the future.

"Dealing with repressed trauma is akin to melting an iceberg."

- The sensitive endeavor of working through repressed trauma is analogous to melting an iceberg.

- Uses include working with patient resistance to dealing with repressed material.

- PTSD

- Defense mechanism

- Patient resistance

For those individuals who bravely choose to resolve significant childhood trauma, achieving healthy resolution requires special care. As an adaptive means, the client, out of necessity, compartmentalized or partitioned off significant chunks of data to keep from being emotionally overwhelmed. This coping mechanism likely prevented them from living in the present with the same affective intensity as the original trauma.

Of course, this partitioning off of experience has led to free floating anxiety and a failure to experience healthy emotions and attachments. This is why they have undertaken this therapeutic endeavor in the first place. This dissociative repression is nothing new to the trained clinician, but to the patient this might be an entirely novel concept. Thus, it is safe to say that a majority of client resistance to dealing with these matters arises out of means in which to protect him or her from significant regression or emotional insult.

It has been my experience that certain therapeutic interventions that lead to immediate catharsis (i.e. confronting the accuser, EMDR, hypnosis or even techniques like rebirthing) can cause potential harm or unprotected regression in the client. It is dangerous and in a way a repeat of the trauma to strip them of their defenses too abruptly. A therapeutic style that I find efficacious in this matter is a gentle, but persistent uncovering approach, demonstrating how current relations and affective experience are influenced by the past trauma. When I sense strong resistance emanating from the client I offer the analogy that dealing with such a sensitive matter is akin to melting an iceberg. There is a necessity to melt this obstacle in order to allow safe passage of more healthy emotions and relationships, but we must be cognizant to not go too fast, and thus risk flooding anything in the iceberg's path.

Usually following the relating of this analogy there is a response of relief and deepening of rapport. Metaphorically then, we can discuss that we can use a small blow torch at first to melt only those parts of the iceberg that one is comfortable with before moving on to larger areas that will require more heat.

Usually the patient comes to realize that they can handle increasingly more emotion and exploration than they originally thought, however I think this is accomplished in part by the therapist demonstrating to the patient their understanding of the necessity to compartmentalize such trauma.

"Upper and Lower case guilt"

- This analogy demonstrates how destructive shame and normal guilt differ by comparing them utilizing uppercase and lower case letters.

- Uses include working with patients raised in shame-based households.

- Ego ideal

- Shame

- Internalized negative cognitive schemas

Helping the patient differentiate between guilt and shame is often an essential treatment goal in the motivating of healthy behavior and self-esteem regulation. Those individuals raised in shame-based family systems have a warped idea of how to feel. This is especially sad in that a healthy dose of guilt has its place in the training of prosocial behavior.

The permissive parenting styles prevalent in the last two decades have almost entirely done away with guilt as a means of creating conscience. Those proponents of guilt-free child rearing argue that it may create shame in the child and thus squelch creativity and lead to low self- esteem. On the contrary, this type of parenting often leads to entitled and narcissistic children who have a very difficult time with delay of gratification.

This swinging of the pendulum toward permissive parenting is understandable. The history of punitive and autocratic parenting steeped in excessive and inappropriate amounts of guilt were previously over-utilized as singular means of motivating behavior. These individuals have a tendency to grow up as perfectionists, with harsh superegos rife with shame based cognitive patterns. The internalized voices of these individuals scream out with guilt of the worst type...SHAME.

It is important to distinguish the two, as shame is an extreme form of guilt much like, for instance, paranoia is an extreme form of cautiousness and perfectionism is an overly developed form of industry. This is especially true for the patient who, due to an upbringing rampant with shame inducing beliefs, has developed a paralyzing internalized questioning of everything they do or choose. In session, these individuals have difficulty distinguishing the relatively normal experience of guilty feelings, having a tendency to morph these ideas into shameful cognitions.

For those individuals it has been helpful to aid them in identifying their experiences in one of two categories. The normal guilt one might feel would be identified as guilt with a lower case g. The unhealthy, and often shame based cognitions emanating from a harsh ego ideal would be identified as shame or guilt with a capital G. Patients who became adept at this analogy would often stop mid-sentence in a session and reply with a smile, "That's a capital G isn't it?"

"Self-flagellation and the internalized voice"

> - This analogy posits that possessing an internalized negative voice is akin to self-flagellation.
>
> - This analogy is often utilized in explaining where self-defeating internal dialogue originates and how it continues to lead to lower self-esteem.
>
> - Cognitive schemas
>
> - Ego-ideal
>
> - Self-negative ideation

The power in psychodynamics of the negative internalized voice of the parent is a hallmark precipitant of psychopathology in later adult life. This patient comes in describing a self-deprecating inner dialogue. They complain of constant feelings of dread and indecision. Their persistent worries include a fear that they will upset or disappoint someone.

A psychosocial history reveals an overtly critical parent. The patient identifies to some degree how this influences their present behavior. However they fail to see the direct connection between their internal dialogue and numerous negative introjects they have suffered at the hands of their often well meaning but psychologically damaging parents. A theoretical explanation of the punitive characteristics of the ego ideal and superego might come across as overly pedantic, but nonetheless the patient must discern that their self-defeating inner dialogue is often a direct result of not meeting their parent's unrealistic expectations.

A visually powerful analogy describing self-flagellation often illustrates how the patient is not in touch with his own true inner voice but instead has adopted negative parental introjects as their own. I can describe that if one is whipped enough as a young person, their own free will is often beaten down. In some bizarre fashion, one begins to feel they deserve it, especially if the one doing the whipping is an individual who they really want to please. In lieu of the absent perpetrator, one has a tendency to develop an internal self-punitive whip.

Therefore, in a sense I explain that they are taking the whip that was once used against them and now doing it themselves. This helps explain to the patient that even though they are no longer exposed to the original source, the same missed expectations still apply. This often allows the patient to better identify these negative schemas and messages.

"The Ego Ideal and how it is similar to the high jump bar."

- In this analogy the ego-ideal, the harsh and punitive precursor to the ego, is compared to setting a high jump bar excessively high.

- Uses include helping patients understand why they are constantly feeling as if they are falling short of expectations.

- Object relations

- Ego-ideal

- Negative automatic thoughts

Patients regularly come into therapy complaining of a life-long struggle with lowered self-esteem and depression. From a cognitive perspective there is always an automatic thought involving never being able to meet the standards of others or themselves. They express that no matter how hard they try they always seem to fall short. This automatic thought rules their existence and they appear to be in a perpetual state of disappointed hopelessness. It is imperative as a clinician to rid the patient of this maladaptive and automatic cognition. The etiology of such a cognitive framework reminds me of the psychoanalytic writer Jacobsen's concept of the ego ideal.

Essentially, he hypothesizes that the child's earliest identifications are rooted in his or her own wishes to become like his or her parents in order to gain parental approval. These early identifications become gradually internalized and show up in the imitation of the parents. These internalizations are quite symbiotic and are the product of weak inner boundaries. The child now continues along the separation-individuation continuum and, with normal development, the child's separation causes a differentiation between self an others. Now the child no longer wishes to be symbiotically one with the parent, but at the same time chooses to emulate certain attributes and characteristics possessed by the parent.

The forerunner to the superego, the ego-ideal now takes precedence. With its development, a beginning representation of internalized moral standards and parental demands are now acquired. Along with this, a budding self-critical aspect of human experience emerges. This ego ideal not only contains internalized moral standards but also material goals and ambitions as well. Thus the child, maintaining such lofty aspirations (internalized parental attributes), begins to try to achieve these with the limited reality constraints placed on a small child. Unable to reach these lofty and unrealistic goals, there is a breakdown in self-esteem.

It is supposed then, that self-esteem depends on the extent to which the individual can accomplish or live up to the standards of his or her ego-ideal. This phenomenon is complicated when the parents are indeed more critical than average, and thus this already ultra-harsh ego-ideal internalizes even more unrealistic standards.

To explain this dense yet critical mechanism for regulating self-esteem, I often compare their situation to a high jump bar being unrealistically placed at too high a height from the beginning. In an almost sadistic fashion, no matter how hard the child jumps, he is just a child and cannot possibly reach this height. Of course, he is told this is possible if he tries harder or gives more effort.

The child, wanting to please the parent continues to go about this impossible task feeling disappointed each time he does not make the height. Of course, it is important to set the bar at a realistic height and expect the child to make goals and different heights as a way of motivating them and helping them internalize a sense of self-competency and industry. The problem in our adult patient's situation is they have internalized a sense of failure and incompetence. Then, as a cognitive behaviorist might posit, they develop a cognitive schema or automatic thought that they would never please themselves or others.

In helping the client break this maladaptive thought pattern and come to some insight as to the origin of this phenomenon, a re-parenting approach is often necessary. This involves the patient setting realistic heights, or goals to achieve. I often reflect in session with the patient whether the bar or goal was set at a realistic height before they automatically view themselves as faulty.

"Sometimes one child acts like a barometer in a dysfunctional family."

- This analogy compares the identified patient in a dysfunctional family to a weather barometer.

- This analogy is useful in explaining family system approaches.

- Family systems

- Identified patient

Children are frequently seen as identified patients in individual psychotherapy, with the implied purpose of the clinician to ostensibly "fix" the acting out child. It often becomes necessary to utilize family systems approaches to exact true healthy psychological functioning.

In taking an initial clinical history, it becomes quickly apparent that the child is lacking any serious clinical psychopathology, but is rather acting out as some response to dysfunctional dynamics. Whatever the acting out, it becomes imperative that the clinician relay to the parent that the system needs to be significantly modified. Simple changes to the child's behavior will be transitory unless the system in which the child resides is changed. As one might expect, this is often a sensitive endeavor indeed. Often these systemic dynamics are quite entrenched, and one or both parents may have to undergo drastic transformations that they are not prepared for. The key to success is determined by the manner in which the therapist approaches this sensitive proposal. Directly confronting the family's dysfunctional system might lead to unnecessary resistance or worse yet, termination. As discussed before, analogies are often an adequate way to approach the client in a distant way until greater therapeutic rapport can be established.

In this rather common example, I educate from a systemic approach, comparing the acting-out child (the identified patient in family systems nomenclature) to a barometer. I explain that a barometer in meteorology is a device that in a sense measures unseen and changing characteristics in the weather and reflects back the results. While an accurate analogy, parents often meet this interpretation with resistance, the parents insisting that the child's behavior is manifested from him and him alone.

The value of a good clinical history comes into play here as the therapist can relate a temporal history using the client's own words about when this behavior first began. A sort of one down approach can be beneficial in this scenario. Something along the lines of; "If the possibility that your child is in some way acting as a barometer, is there anything that may be have been happening in his environment right around the time the behavior started?"

"An air pump is similar to the mechanism that regulates Self-Esteem."

- This analogy compares an air pump and how it fills a slow leaking ball to how one regulates self-esteem.

- Self-esteem regulation

- Sexual abuse survivors

- Sexual promiscuity

This analogy was used with the following client. An attractive woman in her early 30's entered into therapy presenting with a chief complaint of frequent unsatisfactory and transient relationships. During the initial intake and social history I surmised that the origin of these failed relationships was rooted in early sexual trauma and emotional neglect. In this common scenario it was easy to infer that the client's present behavior was at least in part resulting from an early fusion of acceptance and sexuality related to her history of sexual abuse. The failure of these relationships had at its core a faulty mechanism for which this woman maintained her self-esteem. In essence, the client failed to believe that she could be loved or accepted without the sexual component, thus allowing her self-esteem to be too heavily dependent on the success of her relationships with men. In addition, the client initially presented in session as dependent and sexually provocative while merely hinting that she became physically involved early in relationships.

With only veiled hints of premature sexual involvement in relationships, I could imagine her vulnerability and resistance if this topic, along with her nonverbal in-session seductiveness, was interpreted. To avoid early discussion regarding the vulnerable aspects of the patient's promiscuity and its direct contribution to relational failures, I offered the analogy of the slow leaking ball and the air-pump to deal with these core issues until rapport was established.

In this analogy, a ball with a leak is compared to the individual with low self-esteem and the air-pump is analogous to the mechanism with which one regulates self-esteem. It is presented in analogy form in hopes that the patient will see the leak as similar to her plummeting self-worth when her partner withdraws affection. Over time she realized that the hole was analogous to the original narcissistic injury and confusion of her earlier sexual trauma. Furthermore, by allowing men to be the sole regulator of her self-esteem (in this case the air-pump), she saw that she had relinquished her own ability to control her sense of worth.

The use of this analogy provided a starting point where the client and therapist could discuss vulnerable material by focusing on the impersonal and safe content inherent in the analogy. This woman began to incorporate what she termed her "own hand-held air-pump." She labeled it as such because while it didn't quite fill her up as fast as male attention, she felt it was self-sustainable and resistant to the withdrawal of others affection. Among those self-esteem boosters developed were increasing same-sex friends, career achievement and learning a foreign language.

Analogy for repetition compulsion (2)

> - Continually choosing a flawed piece of marble in which to sculpt is analogous to continually choosing a flawed mate in which to engage in a relationship.

Repetition compulsion is that psychodynamic phenomena that explains why people tend to repeat destructive and often tragic relationships. These individuals never understand why they "chase" after the same types of men/women, only to find themselves bitterly disappointed in the end. Essentially, the repetition compulsion arises out of an unfulfilled need to fix or make right the damaged primary relationship one may have experienced with a mother or father. In these scenarios, the father may have been the rogue and charming cavalier who, despite being ultimately irresponsible, may have made his daughter feel special. At a critical time the father may have even left and abandoned the child. That young girl then might begin a life long quest to restore this 'special' relationship, continually seeking out those men, who while exciting, are ultimately abandoning philanderers. This is an unconscious process, the woman not realizing why she continually searches for these ultimately unrewarding relationships. Even with an ample supply of adoring and willing men from which to choose, these appropriate choices are seen as not having the same emotional charge or "chemistry". This is because being with these men will never replace the deficient primary relationship. The repetition compulsion essentially posits that the immature primary process believes that the primary relationship cannot be repaired until the woman can "fix" the current one. Consciously understanding this phenomenon is essential to helping the client break the pattern of destructive and unfulfilling relationships.

For women, I compare the need to sculpt a flawed piece of marble, while a more solid specimen exists close by. For men, I often compare continually choosing a high maintenance roadster when more reliable transportation is readily available.

"Comparing a dysfunctional family to a broken down yet moving car"

- This analogy compares a dysfunctional family to a deficient yet still running car.

- This analogy is useful for families resistant to making difficult changes to their family system.

- Family systems

- Codependence.

Picture this. A beat up 1964 Plymouth Valiant sputtering down the road, bumpers held on by duct tape and wire, smoke from burning oil spewing from the knocking engine as it makes its way down the highway 10 miles per hour below the speed limit. The power steering leaks, so the driver must stop on a daily basis at the local Pep Boys so he can purchase power steering fluid at 2 dollars a pop for the days drive. Brakes shot, the driver compensates by traveling at a much lower rate of speed, hanging way back to avoid the possibility of rear-ending someone. Oil leaks require weekly additive, again at a nominal, yet accumulating, cost. The deficient engine racks up unforeseen fuel costs. When the driver is confronted as to why he continues to operate and maintain this obviously costly heap instead of actually fixing the problem he sheepishly states, "Well it gets me down the road." Queried if this daily cost and foreboding trouble stresses him he retorts, "Of course it does but what am I going to do?"

I often utilize this analogy to explain the seemingly counter-intuitive resolve to remain in or maintain a dysfunctional family. As we all know, change often brings initial conflict and the fear of the unknown. In order to exact change, long-standing and maladaptive dynamics will need to be painfully confronted and resolved. There is a true risk that family relationships might permanently be altered. Resentments and emotional upheaval are likely. The amount of hard work required overhauling a defective engine or family is quite an endeavor indeed. It is important to point out to the patient, much like it is to the car owner, that the daily costs of maintaining these dysfunctional dynamics is nominal. However over time there is an insidious accumulated cost that will almost certainly result in the inevitable catastrophic breakdown of the family.

Analogies of Forgiveness

Many of our clients come to therapy displaying emotional deficits resulting from posttraumatic effects of abuse by a family member. Whether it was physical, emotional, or sexual abuse the client throughout the course of therapy is forced to reconcile in some manner with the perpetrator of said abuse. How one accomplishes this is a matter of debate and clinical training. However, somewhere in the process forgiveness is usually part of the equation.

There is much clinical debate as to whether a client even needs to confront or forgive the perpetrator in order for the patient to reclaim mental health. My best clinical judgment suggests that some sort of forgiveness is probably necessary; however certain prerequisite steps are paramount. Most victims of abuse, as one might expect, are on the passive side and are often too quick to forgive sometimes horrendous abuse. These individuals have difficulty with the expression of emotions, especially anger and often ostensibly forgive just to go on with their passive coping mechanisms. This leaves the patient still feeling angry and frequently, they continue to express it in self-damaging ways. Told that forgiveness will liberate them, they still feel unsatisfied and powerless. This is especially evident in Christian clientele when they are taught only about the passive Jesus skipping over His righteous anger. Following are a couple of analogies that illustrate the prerequisite steps of anger that should precede any attempt at true forgiveness.

"Jesus and the Temple"

- Expressing righteous anger as a perquisite for forgiveness is analogous to Jesus trashing the moneychanger's table outside the temple.

- This analogy is useful in working with patients attempting to forgive.

- Abuse survivors

In the New Testament (Matthew 21: 12-13), Jesus physically manifests righteous indignation outside the temple, angry with the moneychangers that are desecrating the church. This verse portrays Jesus quite differently from the usual benevolent Savior. Righteously angry, he overturns the tables and furiously drives out the charlatans. Being God it is safe to assume that he eventually forgave those who faithfully repented, but not before displaying validated anger.   This analogy effectively demonstrates the need for proper expression of anger before any realistic and successful forgiveness can take place.

"Howard Roark and egoism"

- This analogy compares the concept of letting go of someone's hold over you to a passage in Ayn Rand's classic book, <u>The Fountainhead.</u>

- This analogy is useful for those patients who cannot let go of another's transgressions towards them.

- Forgiveness

In Ayn Rand's seminal novel <u>The Fountainhead,</u> she uses the main character architect Howard Roark, to demonstrate the epitome of her concept of egoism. Among other characteristics Rand portrays Roark as a progressive visionary that is true to his principles and individuality. While not the purpose of her book, I have nonetheless utilized some of its principles in helping patients resolve issues regarding being transgressed against. Rand depicts Roark as similar to Frank Lloyd Wright. In the book, he is commissioned to build a church, which he does in his progressive yet often controversial style. His archenemy in the book a jealous, devious, and less talented architectural critic, summarily criticizes his church in his column. He creates such a stir that the individual who originally commissioned the project gets another architect to makeover the church.

In a key part of the book, both Howard Roark and the critic contemplate this dilemma. The critic, giddy with power and envy, cannot wait to view the overdone monstrosity of a church. Howard, true to his principles, cares little about acceptance or accolades but only what they have done to his vision. He forces himself to view the church. Utilizing literary license both men inspect the church at approximately the same time. They then happen to cross paths where the critic, practically salivating with Machiavellian glee, says to Howard, "Well what do you think of me now?" I usually let the client try to guess the stoic architect's reply. After incorrectly guessing several times, I finally relay what the character said, a perfect example of taking all power from the abuser. He simply states, "I don't think of you". This perhaps idealistic response certainly does illustrate another type of forgiveness. This type allows the patient to forgive himself for no longer being prisoner to whatever suffering he may have self-inflicted as a result of undergoing another's abuse.

Analogies for Depression and Psychotropic Compliance

As we know from our clinical training, the distinction between transient mood states and clinical mood disorders is essential. We often find ourselves having difficulty explaining this keen distinction to our clientele. This is important, as their recovery from such significant clinical disorders often depends on their understanding of this difference, as well as the clinical protocol needed to adequately treat them (i.e. medication compliance, lifestyle changes, etc.).

This confusion is understandable. Our patients often experience a variety of moods; happiness, anger, and sadness, to name just a few. Unpleasant moods and changes in affect are normal reactions in everyday life, and we can often identify the antecedent events or stressors that caused their mood to change. However, when they experience extreme changes of mood that are out of proportion to their normal every day reactions, that seemingly come "out of nowhere" it makes it hard for them to function. These changes are often the result of a *mood disorder*. Many of our patients have trouble with this concept, choosing to forge ahead with their regular methods of stress reduction and mood elevation that have become gradually less effective over time. They often let this continue for too long before finally consulting a professional.

During this initial diagnostic phase of therapy, it is important to educate the patient in respect to differentiating everyday mood fluctuation from the more significant clinical disorders. They need to understand that mood disorders are often biological illnesses, which after a prolonged period change brain chemistry. They need to understand that they are not really at fault for their entire presenting condition, nor are their symptoms the product of a "weak" or unstable mind. Rather, mood disorders are treatable with a combination of therapy and sometimes psychotropic intervention. This is, as we clinicians know, often harder than it seems. These next set of analogies are intended to offer suggestions on how to educate the client, breaking through their natural and prideful resistance, and thus improving treatment compliance.

"Pulling back the troops for reinforcements is often necessary before the next attack."

- This analogy compares the necessity of pulling back tired troops to wait for reinforcements to the need to temporarily back away from an overwhelming stressor.

- Useful when working with depressed patients having difficulty asking for assistance.

- Depression

Many times a strong and mentally capable individual finds himself or herself up against a formidable task they must overcome. Equipped with ample psychological resources, they go about the challenge expecting, as they have most of their lives, to succeed. One example that comes to mind is that of a middle aged married woman faced with caring for her mother, recently diagnosed with cancer. The doctors give the patient six months to a year to live. Consequently, the daughter goes about making these last precious months as comfortable as she can for her terminally ill mother. She runs her back and forth to her chemotherapy appointments, and meets with financial planners, all the while performing her own job and taking care of her own thriving family. The stress is intense yet manageable, and nothing is going to detour her from doing everything in her power to make her mother's last months as comfortable as possible.

The daughter, although blessed with a repertoire of multiple psychological resources, soon finds herself worn down and depleted. Her mother survives into her second year, and the promised help from her siblings never materializes. She begins to develop symptoms of anxiety, depression and irritability, even frequently becoming angry with her mother. Her sense of guilt and duty pushes her on even though she feels she is unraveling from the inside. She enters therapy after feeling guilty for having wishful thoughts about her mother's death as well as frequent arguments with her husband. She is initially resistant to suggestions that she back off and perhaps delegate to other siblings or to consider professional assisted living. She feels an urgent pull tinged with guilt.

I explain her situation by comparing it to a troop of soldiers who are asked to take a fort in battle. They have ardently fought to establish their position just outside the fort. They have lost many men but their goal is close at hand. The battle has lasted far longer than expected and the troops are tired and depleted. They are so close, but each attempt the weary soldiers make on the fort is thwarted with great force and the troops are pushed back, sustaining even more casualties. Orders come from headquarters demanding they retreat to replenish their forces. The superior officer meets these orders with resistance and stubborn reluctance. To pull back now may signify defeat and render worthless all their hard work and men lost. The objective commanders point out, to no avail, that to keep rushing the fort with depleted forces only leads to further losses and a lesser chance of achieving their goal.

I ask the client to discuss the difficulties faced by the superior officer, what motivations would keep him continuing forward, and some possible solutions. This analogy often allows the client to view their situation from an objective position, realizing the need for personal replenishment, and/or added assistance.

"Depression is analogous to a cold turning into pneumonia."

- This analogy compares delaying treating early symptoms of depression to an untreated cold possibly turning to pneumonia.

- This analogy can be useful in normalizing clinical depression for our patients.

- Depression

At times, an individual will enter therapy in crisis, exhibiting obvious symptoms of clinical depression. They describe the usual constellation of symptoms (i.e. difficulty concentrating, ruminative thoughts, early morning awakening, becoming tearful without provocation and overall vegetative malaise). Despite the individual's obvious crisis, the first mention of the words "clinical depression" results in the patient's very denial of the just detailed symptoms. They simply believe this is not possible. "Clinical depression; that's something mentally weak people have. I am just a little stressed, that's all." Many times the client's support system supports this notion, feeling that their spouse/child/friend simply has to 'buck up', pray harder, or simply leave the jerk.

This is easy to understand as we all experience transient mood states characterized by many of the symptoms included in a major depressive episode. However, as is the transient nature of these mood states, they usually dissipate after a short time. The individual entering therapy in crisis has usually expected this mood to pass and is surprised that it has not. In order to maximize patient compliance to a more intensive treatment approach, which may include psychotropic intervention, it is important to educate the patient about the physical manifestations of clinical depression. I often compare the difference between a transient mood state with depressive affect and a more serious depressive episode to how a cold might turn into pneumonia if not treated.

"A car getting further and further stuck in the mud."

- This analogy compares transient mood states turning into clinical depression to a car getting further stuck in the mud.

- Useful in helping clients consider psychotropic intervention.

- Depression

- Medication

I often utilize this analogy in cases where there is not only a need to describe the difference between an unpleasant mood and a more serious major depressive episode, but also when psychotropic intervention is warranted. This is usually after more standard therapeutic methods have been exhausted and symptom relief, despite strong patient efforts, has been minimal. It is often useful for individuals who are prone to be resistant to the idea of medication as a possible treatment method.

I describe that when our automobile becomes stuck in the mud, our first tendency is to step on the accelerator. Finding that this intuitive response only makes us increasingly stuck, we follow our next instinct, which, of course, is to tromp on the accelerator even more. Now further mired in the mud, we do not give up but repeat the accelerator action until we have realized that our wheels are simply spinning. Only then do we contemplate another method of getting unstuck.

Usually we find some sort of wedge, perhaps a plank of wood with which we can at least gain enough traction to propel us from our current state. The wedge doesn't necessarily cure our problem, nor will we rely on it forever, but it may get us on our way where we can better negotiate the murky terrain until we reach more substantial terra firma. In this simile, stomping on the accelerator is akin to the usual methods of mood elevation and stress reduction the patient utilizes effectively to deal with transient unpleasant mood states. It may also represent the efforts of psychotherapy alone. The wedge is analogous to the possible anti-depressant medication. It is important to describe the wedge as a temporary boost, and not the "total" solution to the problem. This assuages some of the fears that go along with taking medication.

"Wading into the mud up to our ankles instead of continuing until it reaches our hips."

- This analogy compares taking anti-depressant medication to deal with depression to walking up to your ankles in mud as opposed to your hips.

- Useful in helping clients to consider anti-depressant medication.

- Depression

Although a majority of my clinical practice involves working with depressed individuals, I often resort to referral for psychotropic medication as a means of last resort. This bias must somehow be counter-transferred to my clients because most are hesitant to take medication for any of their mental health issues.

Their trepidation is understandable. Between the sensationalized horror stories reported in the media and the very real difficult side-effect profile, ingesting a foreign substance to change our biochemistry is indeed a choice that needs to be seriously considered. Well-meaning psychiatrists and physicians often complicate matters with their truncated and solely medical explanations. This leaves us as clinicians to guide our patients through such decisions.

Many of my clients are afraid of the side effects. Some are afraid to give up their autonomy in their own mission to regain their premorbid psychological functioning. They complain that they do not want to take a "pill" that will make them artificially happy, while neglecting the real origin of their depression or anxiety (the Don't-Worry- Be-Happy syndrome). I have had many reports of individuals accurately complaining that their medication dulled many of their real transient sad feelings. Many are afraid that if they start taking medication, they will "never get off it." A lesser group naively believes that their symptoms will magically disappear with medication and there is no need to change environmental stressors.

To more accurately describe the effects of medication in less medical terms, and hopefully assuage their major concerns, I offer an analogy comparing walking up to your hips in mud vs. walking up to your ankles.

I describe that a full-blown mood disorder is akin to walking up to your hips in mud. Despite ones best efforts, this arduous process quickly becomes exhausting, further depleting motivation to continue. I describe that for the most part, taking anti-depressant medication often controls the major symptoms thus allowing the individual to feel as if they are walking only up to their ankles in mud. There is still work to be done, (changing behaviors, altering environmental stressors, and understanding psychological antecedents) however now walking up to your ankles in mud seems much less daunting and exhausting than walking in mud up to your hips. This also quickens the journey out of the metaphorical mud altogether. This latter concept often increases individuals understanding that medication is not a permanent answer but a temporary boost.

"Inertia feeds the monster:
movement starves it."

- Inertia is analogous to feeding the monster of depression and movement is the same as starving it.

- This analogy helps provide a visual image for individuals struggling with depression.

- Depression

Study after study reports the single best treatment for clinical depression is restoration of sleep patterns and exercise. However, if one is clinically depressed, it is extremely difficult to engage in any form of exercise. Nonetheless, it is paramount to get our patients to engage in some form of movement and social activity. They will of course complain saying they do not feel like being around anyone or that they cannot get going.

However, more often than not they do feel at least temporary relief once they accomplish some sort of movement or social interaction. Not doing anything only feeds the cognitive distortions of the mood disorder.

In order to emphasize this phenomenon, comparing clinical depression to a grotesque monster is often helpful. After having them envision the most hideous creature imaginable, I ask them to visualize it slovenly eating in gluttonous fashion. I then tell them that inertia, or the absence of movement, feeds this creature and movement starves it. Sometimes I ask them to visualize the agonizing death of this disgusting form as they exercise.

"Explanation of rebound effect following stopping medication"

```
• This analogy compares rebound effect
  following cessation of psychotropic medication
  to employees needing to increase their
  workload once the temporary help leaves.

• This analogy is useful in preparing patients
  with discontinuing psychotropic medication.

• Anti-depressant medication
```

Invariably as one titrates down from anti-depression medication, there is a slight rebound effect. Individuals abruptly taking themselves off their dosage often pronounce this phenomenon. By rebound effect, I mean that feeling of transient return of depressive symptoms. To better explain this effect absent of physiological jargon (which I have a hard time doing anyway) I ask the client to imagine Serotonin production as akin to factory workers. In this case I ask them to imagine that these workers, once exhausted in their ability to produce the mood enhancing chemical, are becoming quite used to the extra boost provided by the SSRI. So comfortable in fact, that they may have become lazy, taking extra or lengthier breaks.. Therefore, when the extra workers are abruptly taken away, the original workers are startled and may take awhile to step up their production. I may ask them to imagine the production workers sitting around leisurely doing nothing when all of the sudden there is an alarm reminding them to get cracking. There is a lag time before they can naturally produce the required amount of serotonin or norepinephrine, thus one feels a minimal recurrence or rebound of their depressive symptoms until production can catch up. This analogy is also helpful when dealing with laxative abuse as a dietary strategy. Once there is a cessation of the laxative, constipation often follows before natural digestive processes can kick in.

"The James Bond ultimate ray of destruction"

- This analogy compares the cumulative power of simultaneous multiple stressors leading to depression to the adjoining rays of destruction in James Bond movies.

- This analogy can be useful in helping clients dealing with multiple stressors; explaining that the lessening of only one stressor may in fact lessen the entire burden.

- Depression

This multi-faceted analogy compares converging events in one's life that sometimes lead to emotionally devastating effects to the omnipresent weapon of destruction in all James Bond movies. You remember the classic and ubiquitous scene where James Bond, caught in the underground secret lair of his adversary, is forced to listen to the explanation of how the world will be destroyed if the million, billion or trillion (depending on the decade or whether the movie stars Connery or Brosnan) dollar ransom goes unpaid. Usually a series of three or four rays are present that when joined or linked together converge to form a powerful ray of mass destruction. Of course, Bond routinely escapes his less than brilliant captors and in the longest countdown imaginable (twenty second sometimes lasting two minutes) he knocks away the last converging ray, rendering the mass destruction ray impotent.

Individuals entering psychotherapy are often emotionally frayed and fragmented, suffering from a clinically significant major depressive episode. This is what the layperson may describe as a "nervous breakdown". They feel they cannot handle even the slightest stress without falling apart, complaining of the proverbial 'end of the rope'. In recounting their social history, it becomes readily apparent they are suffering the deleterious effects of multiple and converging stressors.

Much like the rays of the Bond films, it is not one individual stressor that has them in such dire straits, but the convergence and accumulation. They are often overwhelmed by such a confluence. Examples of this could be the loss of a loved one, a divorce or being the victim of corporate downsizing. These factors, singly or in conjunction, leaves them depleted of their psychological resources. Explaining that we only have to ameliorate one of the stressors to render the overall effect impotent helps them compartmentalize these 'rays' and focus energy on those stressors that are in their control.

This analogy is also helpful at the end of treatment where a major depressive episode has been adequately treated. Many times during termination, the patient worries that with the cessation of therapy or psychotropic treatment they will regress to that overwhelmed emotional state. Occasionally it is important to review and indicate that the chance of all those stressors converging again in the same way is quite rare, plus they now have a better understanding of what role those stressors play in their lives.

"The process of battling depression is akin to rolling a boulder up a hill."

- This analogy compares the initial steps of battling depression as akin to rolling a boulder up hill.

- This analogy is useful in helping patients suffering with depression to realize the first steps toward treatment are often the most difficult.

- Depression

- CBT

In describing the process of battling one's depression during the initial phase of treatment, the therapist must guard against making the process too simple. Our clients before finally deciding to seek professional assistance, have surely endured constant barrages from loved ones to simply, "Buck up", "Just change your diet", or "What I did was…." They have drained all of their reasonable strategies to rid themselves of these persistent symptoms and need badly for someone to understand their impasse. As clinicians we walk a fine line, for while our patients are searching for a cure to their troubles, still they want us to understand the amount of pain and malaise they are feeling. In an attempt to walk this fine line, I offer the following analogy of negotiating a major depressive episode.

I compare the journey of battling depression to walking up a hill pushing a large boulder. However, I assure them that in most of my clinical cases, if proper treatment protocol is complied with, the boulder tends to get smaller as they walk up hill and there is usually a summit to this episode. As they reach this point, the hike becomes downhill and the once substantial boulder has now been reduced to a couple of five pound rocks. Thus, with this analogy I emphasize that starting is the crucial first step. I mention that by coming in they have already started the process.

"The smoking computer"

- In this analogy, excessive ruminations are akin to smoking computers in old movies after receiving a question in which there is no good answer.

- This comparison is useful in demonstrating the tautological and deleterious effect of constant ruminations.

- CBT

One of the significant factors leading to the onset of any clinical depressive or anxiety episode usually involves excessive rumination. The mind becomes overwhelmed with some unexpected piece of information, which despite any and all type of cognitive analysis renders only more questions. I'm talking here of the real bombshells: being told one is the father of a 2 year old daughter, being laid off from a job one has had for 20 years, the discovery that their spouse has been involved in an 8 month affair. All of these change the way one views the predictably of their environment.

The systematic cognitive schema of the brain, ever searching to maintain equilibrium and predictability, frantically processes the information in an attempt to assimilate the new information to an existing schema. Unable to initially achieve this, the mind does not immediately attempt to form a new schema but continues to try and fit it into an old pattern, leaving the individual exhausted as the accelerated processing continues. The patient feels out of control of this process, as each time they try to make sense of this overwhelming dilemma, the answer is unsatisfying.

We see this manifested in therapy by the client who continues to utilize session time trying to comprehend why someone or something could place them in this situation. They want us to make sense of this data so they can understand. We see this process rendering them exhausted and sleep deprived, leaving them vulnerable to a more clinical depressive episode. Objectively we can see the need for the patient to form a new and synthesized worldview or cognitive schema, painstakingly evacuating the old understanding of their environment. It is difficult to breakthrough their resistance because each attempt to offer a reframed perspective is met with an almost automatic response that rigidly holds fast to an already existing schema.

A fitting analogy seems to be one of comparing their dilemma to that of feeding a computer a question that has no rational answer. I have them visualize those scenes in an old movie where the refrigerator size (remember those?) computer is given a question it cannot answer. As they usually recall, the tapes on the computer churn with accelerating speed until they begin to smoke and the computer eventually breaks down. That vivid image is usually a turning point in the patient seeing the necessity of attempting to control the obsessive rumination through implementing some form of cognitive behavioral strategy or medication to minimize this rapid and often tautological processing.

Analogies for Anger Management

"Expression of anger is akin to releasing liquid contents under pressure."

- This analogy compares the individual's ability to express anger to the releasing of carbonated soda under varying degrees of pressure.

- This analogy is useful with individuals displaying over-controlled hostility.

- Overcontrolled hostility

- Abuse victims

- Intermittent explosive disorder

Many times the emotional origin of the presenting conflict in psychotherapy is simply too great for the client to deal with in the early stages of treatment. Analogies allow for the maintenance of an objective perspective until a greater level of interpretation can take place.

A client enters therapy complaining of intermittent and explosive episodes of rage and anger. During the intake sessions the clinician discovers that the individual was subjected to frequent physical abuse as a child. There is a distinct correlation between the early physical trauma and the patient's present manifestation of rage and explosiveness. It also becomes clear early in therapy that the patient's defense against this terrible insult is to present on the surface an air of imperturbability and macho bravado. This individual is the definition of over-controlled hostility. Simply interpreting his behavior and stoic presentation as a defense protecting a hurt little boy during the early phase of therapy runs the risk of the patient experiencing agitated indifference, early termination, or worse yet, regression with no real set of adaptive defense mechanisms to replace the ones he has. It would seem imperative in this case that the client is able to maintain a cognitive or objective perspective regarding his core dilemma until he is more comfortable talking about his early abuse. In this situation, the analogy of the shaken pop bottle can aid the client in maintaining a cognitive distance from the extreme emotional material while at the same time allowing him to work on therapeutic goals of reducing the frequency and intensity of his rage.

This analogy compares the individual's ability to express anger to the contents of a soda bottle under varying degrees of pressure. In session, the therapist may relay that individuals dealing with anger issues often resemble an unopened carbonated soda bottle. Each shake of the bottle, no matter how slight, causes the pressure of the contents to increase. If there is not a corresponding release to each increase in pressure, there is an accumulation that can explode after a time, even with just a slight lifting of the lid. Complicating matters, it seems impossible to tell from looking just how much the unopened bottle has been disturbed. The liquid appears still, regardless of the amount of times it has been shaken.

As one can see, the analogy contains many of the dynamics a sufferer of intermittent explosive episodes deals with. The accumulation of significant, as well as every day, stressors is seen as analogous to the shakes of the bottle. The inability to possess more subtle and appropriate ways to express mediated doses of anger is in this case similar to not releasing pressure by periodically lifting the lid. In addition, the false appearance of the undisturbed liquid is synonymous to the imperturbable defense these individuals often present. Moreover, the opening of the much shaken bottle leading to the volcanic eruption of your favorite carbonated beverage is analogous to the violent display of rage after sometimes only the slightest provocation.

Using this analogy, treatment strategies begin to emerge that can be rationally discussed and implemented while simultaneously building rapport where later discussion and interpretations of early trauma take place. Treatment goals can focus on helping the individual recognize and keep inventory of the various stressors, or 'shakes of the bottle.' Once the stressors are identified, the therapist can encourage the client to develop methods of periodically relieving pressure (i.e. exercising, practicing a hobby, relaxation). Treatment goals can also focus on the training of mediated methods of expressing anger. For example, the use of assertiveness training works well for those situations when the 'pressurized' bottle wants to explode. Perhaps only when the patient recognizes that they have reached a plateau in their therapeutic gains, as one of my patients recently did, is their therapy ripe for deeper discussion of historical material. "Dr. Seaton. No matter how much I try and relieve this pressure, it seems I'm always under pressure." It was at this time that I was able to venture a historical interpretation regarding the connection between his early childhood experiences and present manifestations of rage.

"Anger management is analogous to a thermostat."

- This analogy compares the recognition and expression of angry feelings to how a thermostat controls the temperature.

- This analogy is useful when helping patients recognize antecedents to more aggressive feelings.

- Overcontrolled hostility

- CBT

Many of the patients I see in my practice for anger management fall under the category of overcontrolled hostility. These individuals usually present with imperturbability often noting that nothing bothers them. However, their explosive anger, seemingly materializing out of nowhere, wreaks havoc in their personal and professional lives. On closer inspection we can usually identify an accumulation of small transgressions built up over time and not acted on in an assertive manner.

I sometimes compare for these patients the way that managing anger resembles the way a thermostat functions on an air conditioner. When the temperature reaches a certain point the thermostat recognizes this, making a click indicating to the condenser to operate in such a way as to reduce the temperature back to normal. If the thermostat fails to recognize the rising temperature, the condenser fails to kick on thus allowing the room to become warmer and warmer. To cool the room back to the desired temperature, the condenser will have to work twice as hard. That is if the thermostat indeed ever triggers the condenser.

Our overcontrolled client's first need is to recognize slights, upsets, and transgressions. In other words, they need to develop a more sensitive emotional thermostat. These individuals often fail to recognize psychological and physiological signals usually associated with emotional upsets. I am speaking here of the clenched jaw, the rise in blood pressure, or tightness in the stomach. Ignoring these signals, the angry client fails to self-direct behaviors designed to bring an emotional equilibrium. These situations are allowed to accumulate and fester until the client reacts in an overly aggressive fashion.

Helping the client to create a more developed emotional thermostat is the initial step in a two-pronged approach to dealing with healthy modulation of anger. Developing an expanding repertoire of skills and directed behaviors (i.e. assertiveness training, exercise, stress management techniques) designed to properly express anger is the second goal. In this analogy the acquired skill set is similar to the condenser on an air-conditioner unit.

"Anger as a propelling or destructive physical force"

- This analogy describes the expression of aggression as either a useful or destructive force depending on how it is harnessed or directed.

- This analogy's uses include working with patients displaying passivity, over-controlled hostility, intermittent rage, and depression.

- Psychodynamic explanation of depression

- Overcontrolled hostility

- Aggression

- Assertiveness Training

In the course of human existence anger plays a definitive force in how we conduct our lives. Many individuals deny anger, internalize it, or express it in a destructive and diffuse manner. The internalization of anger is a critical component of the classic psychodynamic explanation of depression. Denial or minimization of aggressive impulses is part and parcel of the overcontrolled hostility personality. In addition, indiscriminant episodes of rage are always a factor in the cycle of abuse. It is important to understand that aggression is a normal and, if harnessed correctly, productive life force propelling the individual forward.

In therapy I often liken aggression to a fuel or a force that has tangible and physical impact on all that it touches. I ask the client to imagine this aggression as a real physical energy. This is a force that if turned inward (in the case of depression), implodes in on the patient resulting in, among other things, the presence of self-negative ideation and lethargy through impotence. For the overtly raging patient, I point out that this powerful force is discharged in a diffuse manner, much like the damaging floods of an out of control river. This same energy, if harnessed through a dam, can create electricity. For the over-controlled individual, I explain that the denial of this force can lead to a dangerous accumulation that tends to come out in an unproductive manner under the least provocation.

Individuals who are in the throes of a breakup or loss discount anger, choosing only to feel the sadness or self-reproach of the situation. I attempt to show that aggression or anger can be adaptively channeled into a positive emotion that can propel the individual. For example, vigorously working out with weights as a way to improve self-image is a healthy means of "getting back" at the other. Assertiveness training is another example of this properly channeled essential life force. In essence, it depends where we direct this rocket fuel. Do we face the engines toward the craft or do we change directions, allowing the force to be discharged in a manner that jets us from where we now find ourselves?

Analogies for Professional Issues

"Tune-up sessions following termination"

- This analogy compares return to psychotherapy after termination to tune-up sessions after significant repairs to a car's engine.

- This analogy is helpful during discussing of therapy termination.

- Therapeutic relationship

- Termination

After a healthy stint of mid-to-long-term psychotherapy, patients are understandably cautious during the termination stage of therapy. Eager to be totally independent and try out the various insights and acquired coping mechanisms gleaned from therapy, they are nonetheless experiencing trepidation at the thought of leaving the safety of the therapeutic relationship. Just as much, they often worry about relapse or a life-event resulting in a return to therapy. Maybe for the first time gaining their autonomous legs underneath them, they are afraid that they may once again totally relapse into the same state that brought them to therapy in the first place. I usually assure them that this rarely is the case, but rather they may return from time to time for a short duration to clarify therapeutic gains.

I liken these short-term arrangements to tune-up sessions in this analogy. Most of the gains they have achieved are permanent and impervious to drastic change. Much like a total engine overhaul, their work has been extensive and life transforming, but they may need a tune-up from time to time.

This has been consistently born out, even when individuals have come back insisting that they will need to again undergo longer-term sessions. Inevitably, they find after only a few sessions that their internalized representations of the therapeutic situation are readily accessible to their working ego, and they are making their own quality interpretations post-haste.

"Major dental work vs. check-ups analogy for kids and therapy termination"

- This analogy compares psychotherapy to major dental work and check-ups.

- This analogy is helpful in explaining the concept of termination in psychotherapy.

- Therapeutic relationship

- Termination

The concept of termination with children is just as important as it with adults. Furthermore, because children lack more stabilized self-representations, the concept of termination can be even more crucial. This is why I sometimes use the analogy of a dentist and sporadic check-ups as a means to explain to the child the need for a possible return to short term therapy.

I make clear that a lot of the work we have done in session to improve a behavioral problem is similar to fixing a real infected tooth that involved a lot of hard work, time, and even a little pain. I go on to explain that they might have to come back for a simple check-up or two so that the tooth will not become infected again. I usually ask them to tell me what they might be feeling or how they might be behaving in order to ask their parents if they can go back to Dr. Seaton for a check-up. Children are surprisingly accurate in identifying those feelings or behaviors that would constitute a need for a return to treatment. I always assure them that I will remember that behavior, or infected tooth, and I will be able to help them fix it again

"Analogy for late charge"

- In this analogy, charging for a missed appointment is compared to an employer notifying an employee they will not be paid for one of the hours in their workday.

- This analogy is useful in explaining the necessity for a charge for missed appointments.

- Therapeutic relationship

- Professional fees

The collection of fees for professional services rendered, an essential part of any successful private practice, is rarely taught in a graduate school setting. It is as if we in the helping professions are almost embarrassed to collect a fee for providing an important service. Altruistic beings that we are we often find it difficult to justify our fee. Nowhere is this more difficult than in explaining our no-show fees for missed appointments. I charge half-fee for sessions missed without 24-hour advance notice: standard, if maybe a bit on the lenient side. However, I once had a valued supervisor who charged full fee for sessions cancelled with less than one week notice (Needless to say, this individual had no trouble justifying collection of fees for his services.)

This was an acquired trait for me, but I now demonstrate much more efficiency. Most clients intuitively understand the need for the charge and offer little resistance. However, some individuals are more than a little brusque when informed that they have to pay for a no-show or a late cancellation. I have found that the following analogy aids me in explaining the need for such a fee without making me sound selfish.

I ask them how they would feel if they arrived one morning to their place of employment and worked a few hours before being informed midday that they would not be paid for the following hour. They would not have to necessarily stay for that hour, but they would have to be back for the next hour to be paid for that one, unless otherwise notified. Any individual who has held an hourly wage job can immediately understand the frustration and inconvenience that this arrangement would entail, and they will usually admit to the fairness of this policy.

"Therapist as navigator"

- This analogy compares the psychotherapist's role in psychotherapy to that of a navigator on the high seas.

- This analogy is of use in setting up expectations during the beginning phases of psychotherapy.

- Theoretical orientation (directive vs. non-directive psychotherapy)

- Therapeutic relationship

There is a delicate balance struck between client and therapist pertaining to how directive the therapist must be vis-à-vis the amount of advice given their patient regarding life choices. There is substantial variance among theoretical schools of thought, but one thing is certain: the client has their own *a priori* assumptions about the process or is unaware of the therapist's bias. It is essential, to avoid unnecessary therapeutic resistance, to clarify these matters during the beginning phases of psychotherapy.

The level of directness may be realistically varied depending on the status of payment or contract (managed care eight session limit vs. open-ended). Still, I choose a school of thought which emphasizes patient empowerment, insight, autonomy, and responsibility as desirable and necessary to the process. The analogy or metaphor I use is therapist as navigator with the client/patient as captain on the open seas. While I won't steer for them, my expertise and knowledge of these waters will aid them in staying on their selected course and will keep them clear of grave dangers. This is accomplished through interpretation of their choices and motivations based on knowledge of human psychology. In addition, I will also consult and aid them in clarifying both positive and negative consequences of their actions. With my expert knowledge of the various psychological illnesses and their severity, I can assure them that I may at times be more emphatic in my navigation so they will not veer too far off course into dangerous waters.

"Hiding slaves: Harriet Tubman's Underground Railroad and managed care"

- This analogy compares protecting our client's delicate psychological information from managed care companies to the hiding of slaves.

- This analogy is useful for the psychotherapist considering their delicate role in working with managed care companies.

- Treatment planning

- Managed care

- Therapeutic relationship

The permanent insurgence of managed care forces the therapist to confront many dilemmas in the therapist-patient relationship. It clouds and obfuscates the boundaries of who the client really is and what information belongs with whom. As we often experience, managed care companies are quite savvy in their attempts to restrict care, requiring the revealing of treatment plans and exclusionary diagnoses, and almost forcing the therapist to label our patients in potentially damaging ways that will carry forward as a preexisting condition. It is a delicate balance to maintain personal integrity, while still providing enough information to provide access to adequate patient care without setting them up for difficulty later on when they change employers (or their employer changes insurance carriers, for that matter).

I have likened this matter, perhaps in overly dramatic fashion, to the dilemma the members of Harriet Tubman's Underground Railroad must have faced as they hid slaves from the law. They were forced to give false information to their interrogators in order to insure the slave's rightful anonymity. Perhaps this is my rationalization or my justification to a higher level of Kohlberg's moral authority when I am at times a bit evasive when it comes to information regarding a more potent diagnoses or what I choose to reveal in my progress notes.

"Which cruise you signed up for is akin to which managed care plan you chose."

- This analogy compares the type of insurance plan our client's have (indemnity, EAP, 6 session benefit) to the varying levels of an ocean cruise.

- This analogy is useful for the therapist in treatment planning with their clients.

- Therapeutic relationship

- Managed care

Another dilemma facing the young therapist is balancing the amount of effort required for the client who enters therapy utilizing their free 6-session EAP benefit as opposed to the motivated patient who is paying our full fee in cash. The altruistic predisposition of most therapists implores us to treat both patients as equal, giving just as much time and energy to both. One quickly realizes that even if we could, the time constraint of the shorter-term therapy precludes us from satisfactorily resolving the client's chief complaints. This is inherently difficult because a majority of the time both patients are motivated and seeking our assistance.

The therapist needs to be aware of this problem because the possibility of ambivalent feelings manifesting themselves in countertransference is inherent when disproportionate fees are involved.

During my doctoral internship, a more experienced supervisor helped me with this dilemma, essentially explaining that the patient or client must shoulder part of the responsibility. I must admit that this went against much of my intuition and training, but after her analogy, it helped me work past pathologizing the patient, a concept not unfamiliar to the neophyte clinician. She explained to me that the client, much like they might sign up for a cruise, has a choice of what insurance plan they choose. Do they sign up for the bargain basement plan that only authorizes catastrophic care, or do they value treatment enough to pay a greater premium? Much like a cruise, does one sign up for the floor that is just above the cargo hold? Or do they spend more to stay in grand rooms with expansive ocean views? Should both of these passengers be treated exactly alike, and do they really unconsciously expect to be? Even if they are wooed by a good sales pitch and sign up for a cruise that is too good to be true and find out it is just that, you can bet they will decipher the fine print next time.

Although this analogy doesn't always translate in my divvying up time depending on plan, (I'm much more likely to treat all individuals in equal fashion), it does aid during those times that I might be dealing with ambivalence that may show up as countertransference. In a practical way, I may consciously give the full fee patient a more desirous appointment time than I would the managed care patient.

"Therapy as akin to a pressure cooker"

> - In this analogy, the inherent boundaries of good psychotherapy are comparable to food cooked in a pressure cooker.
>
> - This comparison is helpful in explaining the need for proper boundaries in relational and group counseling.
>
> - Therapeutic relationship

Discussing too much of in-session work outside of session can dilute or waylay much of the effectiveness of group and conjoint counseling. Not keeping appointments or scheduling them too far apart also affects the continuity of individual therapy. To explain this requirement to groups and couples I often compare psychotherapy to a pressure cooker. We might want to isolate the ingredients and place them in an environment under high and intense temperature. This leads to faster cooking, or in the case of psychotherapy, faster treatment progress. It also seals in the authentic flavor, being less prone to dilution or contamination of other elements.

This type of approach also speeds up the mixing of the ingredients or issues, avoiding the "we are always on our best behavior in here" phenomenon. In individual psychotherapy, this analogy can be used to emphasize the importance of consistent and closely spaced appointments.

Analogies for Child Therapy

The evolving nature of cognitive development in children often precludes the understanding and communicating of abstract thought usually associated with traditional psychotherapy. This precept, along with the social awkwardness and concomitant anxiety inherent in communicating with an adult about their private feelings in a clinical setting, underscores the need for a transitional language such as analogies.

The development of specific therapies for children (i.e. play therapy or art therapy) are less dependent on traditional language and frequently utilize analogies and metaphors as a means to help a child gain insight into their psychological functioning. Depending on where the child is along the developmental continuum dictates the complexity and type of analogy a client might offer. To better understand this concept, a brief review of Jean Piaget's cognitive developmental stages is in order.

Piaget of course posited that the brain is a constantly evolving mechanism that seeks to perform increasingly complex operations through the assimilation and accommodations of new information into existing cognitive patterns. For our purposes we will focus on his stages of preoperational (children ages 3-6), concrete operational (ages usually 7-11 yrs), and formal operations (adolescents).

Preoperational age children are usually only recently past the initial sensory stages of development and can probably only make very simple categorizations, understanding mostly simple differences and similarities between events and objects. Much of their insight in therapy will come from mastery of situations resulting from repetitive play symbolizing a particular trauma. Even simple linguistic analogies may be too complex to be of much therapeutic efficacy.

However, by the time a child reaches the ages between 7 and 11, the cognitive structures developing during this time enable the child to experience an enhanced sense of the environment. Piaget labeled this stage of cognitive development as concrete operations. The most important aspect of this stage is the child's emerging ability to conserve operations. He or she is now able to simultaneously hold more than one operation at a time in his or her mind. A fairly stable internal representation of both self and others has developed during this time, allowing the child to compare similarities and differences to others in respect to a myriad of concepts. They can make categorizations of concrete events for the first time during this stage of development. They are much more amenable to the offering of analogies that are concrete in nature: simple "if then" comparisons, if you will.

By the time a child a child reaches adolescence, they are able to participate in a rudimentary form of abstract thought similar to adults. Piaget labeled this stage of development formal operations, which allows the child to entertain "as if" thinking. He is able to postulate about the future and conduct higher abstractions. Analogies are likely to be able to be more abstract in nature, but the clinician must be aware that due to the adolescent's newly acquired capability, they may over generalize, personalize and distort reality.

"Divorce is like going through a storm."

- A child going through a divorce is akin to weathering a storm.

- This analogy is useful in play therapy situations where a child is dealing with divorce.

- Play therapy

- Children and divorce

One example of transferring the symbolic nature of play to a basic linguistic insight involved a play therapy client in the midst of his parent's divorce. A 6-year-old boy of slightly below average language expression was brought to my office due to externalized aggressive acting-out both in class and at his after school day care. Clinical history revealed that the parents were in the midst of a tumultuous divorce, and my client was exposed to elevated levels of domestic discord, mostly in the form of heated verbal exchanges.

As is the basic theory behind many schools of play therapy, a child will seek to master their situation through the means of repetitive and symbolic play. My patient was no different. He utilized the sand tray to build elaborate structures around two boy figurines, one older and one younger. I interpreted this as more than a coincidence because the patient also had a younger brother. Almost immediately after he had built the structure, he would imitate a violent storm that would destroy the protective shelter. This could be easily dismissed as basic male set 'em up and knock 'em down play, but he did this session after session.

My supervisor at the time, Kevin O' Connor, a noted play therapist, validated my hypothesis that the storm was probably the over-determined amalgamation of the consequences of the divorce. He suggested that I transition the play from inside the sand tray to more of a participative play where we would build structures out of materials in the office to shield ourselves from the impending storms.

We did just that, but not before my patient misinterpreted my suggestion, initially attempting to climb into the small sand tray, inviting me along with him. After a session or two spent behind a makeshift shelter of cardboard boxes weathering a succession of violent storms, I finally made the historic and analogous interpretation linking the storms to his personal situation. "Wow this storm is pretty scary!" I said "This is probably how it feels when your parents are fighting. Just like a bad storm, and no matter how you try to protect yourself your parents still argue." My patient immediately had tears form in his eyes as I told him we were going have to try to fix this. We did come up with some helpful solutions, including telling his parents he did not like it when they argued. We also discussed what happens when a storm comes. He correctly answered that it always ends.

Soon after this one session the content of his play dramatically changed. He abruptly shifted from the repetitive shelter building, to haphazardly playing with the various toys that had always been present. This seemed to be a signal that he had mastered what was his chief complaint. There had been a moderate reduction in his behavioral problems at school, and we soon terminated treatment.

"A psychologist is like a "feelings doctor".

- A psychotherapist is compared to a "feelings doctor".

- This analogy can be useful in helping a child explain what you do in psychotherapy.

- Treatment planning

- Therapeutic relationship

- Play therapy

I am not aware of any research regarding what a young child's (ages 3-7) *a priori* assumptions are about what a psychologist is or does. However, it is my belief that the more a child knows about what we do and what treatment goals are positively correlates to efficacious therapeutic outcome. What does a parent tell a young child prior to his first visit to a psychotherapist? How does a psychologist describe what he or she is during the first session? An analogy whose origin now escapes me is to compare to some extent what we do as akin to a "feelings doctor".

In choosing to compare what we do to the medical physician, there seems to be many commonalities. Similar to Piagetian concepts of assimilation and accommodation, the child will likely compare the common characteristics shared by both professions (i.e. authority, caretaking, etc.).   It is safe to assume that at least a portion of our child clients already comprehend this. I have had one startling experience where a patient's 3-year old sibling accompanied him to his therapy visit and immediately upon entering my office covered his ears and shrieked with terror. Upon inquiring of the unsurprised mother she informed me that the child had experienced frequent ear infections and expected that I would continue a series of intrusive examinations. This child's brother, my client, had been told the following analogy; he, unfortunately, had not.

I usually instruct the parents to relay to the children that they are going to see a feelings doctor. The script is along the lines of: "It is like when you are sick, or have something hurting you go to a doctor and he or she does something to make it feel better. Well, Dr. Seaton is kind of like that, but he helps people who are having some sort of bad feelings, like maybe they are mad too much or cry too much, or worry too much.". I usually reiterate this analogy, and tell them that instead of using shots or medicine to help them have better feelings, we will pick a behavior to change and talk about it or think of ways to play to help make this feeling better. Even with kids in play therapy, I think it is important to have an implied behavioral contract, so that the child has some investment in what his or her play might represent, and how it is differentiated from non-therapeutic play.

"Stuffing your feelings is like stuffing cotton into a medicine bottle."

- Stuffing your feelings is akin to stuffing cotton into a bottle. Eventually they will spill out.

- This analogy demonstrates in visual manner for latency age children how storing feelings can be detrimental.

- Child therapy

- Passively dealing with feelings

My girlfriend has had wonderful experiences utilizing this analogy with her inner city latency-age clientele. Furthermore, it seems to fit the classical sense of concrete operations. Many times, we see in our offices the resistant latency-age child. Already reticent about exposing feelings and internal thoughts, they may want to share but lack the articulation capabilities. Instead of bombarding the child with intrusive questions, which only serve to entrench the child even more, she simply retrieves an empty pain reliever bottle and some collected cotton balls from previous pain reliever packages. (Gee, I hope it's not me that causes her to go through so much Ibuprofen.)

She demonstrates to the quiet but obviously attentive child the need to deal with pent-up feelings by stuffing the very small and light pieces of cotton into the bottle. After insertion of each piece of cotton, she closes the bottle tightly. Then she opens the bottle again to stuff another cotton ball, telling the child that this represents the stuffing of feelings: each one seems light and insignificant but after a while they build up. She demonstrates that after enough time your built up feelings... (She opens the crammed-full bottle and the cotton balls simply pop out and spill over)... need some form of release. While not always immediately acknowledged, it has left a lasting visual analogy that the child can refer to later in therapy.

"Comparing child support to a pie chart"

- This analogy compares paying child support to a pie chart.

- This analogy offers a visual explanation to a child how child support is formulated and how both parents contribute.

- Child therapy

- Children and divorce

- Splitting

One of the more sensitive yet pertinent aspects of dealing with the concrete operational child in a divorce situation is explaining the division of child support. As often is the case, the child, privy to excessive parental information, has formulated an inaccurate opinion regarding the dispensing of monies to raise him and his siblings.

The parent paying the bulk of the percentage, fueled by bitter resentment and hostility toward the primary custodial parent, may slip up by conveying such phrases to the children: "I can't afford to do much with you kids this weekend because I have to send all my money to your mother." These statements, meant to gain sympathy from the child and thus further an alliance, often have deleterious effects. I have observed children so worried about finances that they turn down dessert or decline Christmas presents, because they inaccurately think their parent is close to being put on the street.

They often develop underlying hostility toward the custodial parent, feeling they are unfairly reaping financial rewards at the expense of the other parent. Because of their mixed loyalties, they are rarely direct about these feelings. Instead they manifest them through internal worrying or indirect acting out in response to their parent's spending habits.

To the child in concrete operational phase of cognitive development, this is a difficult concept to grasp indeed. True to form, they see things in concrete terms: the one parent pays x number of dollars while the other parent ostensibly receives it all while paying nothing. The adolescent can abstract that both parents pay, but many times are self-absorbed, caring only that someone foots the bill. To the middle-age child, this is confusing. To alleviate some of this confusion and anxiety, I utilize a pie chart during session in hopes of clarifying in concrete terms where the money goes and that it is a *shared* responsibility between parents.

First, I explain that the court has a formula that decides how much the total cost is to raise a child, emphasizing that the court thinks its fair that the total be the same as if both parents were still married. After all, it was not the child's fault that the parents filed for divorce. We then draw a large pie chart where we identify all the things it takes to raise a child: camps, utilities, food, clothes, and so on. Second, we come up with the approximate time the child spends with each parent. Following the identification of these visual and concrete variables, it is then much easier to demonstrate to the child how the non-custodial parent's child support is used to pay for everyday things the child uses.